Metal Matrix Composites

Metal Matrix Composites

Special Issue Editor

Manoj Gupta

MDPI • Basel • Beijing • Wuhan • Barcelona • Belgrade • Manchester • Tokyo • Cluj • Tianjin

Special Issue Editor
Manoj Gupta
Materials Group
Department of Mechanical Engineering
National University of Singapore
Singapore

Editorial Office
MDPI
St. Alban-Anlage 66
4052 Basel, Switzerland

This is a reprint of articles from the Special Issue published online in the open access journal *Applied Sciences* (ISSN 2076-3417) (available at: https://www.mdpi.com/journal/applsci/special_issues/Metal_Matrix_Composites).

For citation purposes, cite each article independently as indicated on the article page online and as indicated below:

LastName, A.A.; LastName, B.B.; LastName, C.C. Article Title. *Journal Name* **Year**, *Article Number*, Page Range.

ISBN 978-3-03936-092-5 (Hbk)
ISBN 978-3-03936-093-2 (PDF)

Contents

About the Special Issue Editor

Manoj Gupta is a former head of the Materials Division of the Mechanical Engineering Department and director designate of the Materials Science and Engineering Initiative at NUS, Singapore. He did his Ph.D. at the University of California, Irvine, USA (1992), and his postdoctoral research at University of Alberta, Canada (1992). In August 2017, he was highlighted as among the top 1% of scientists in the world by the Universal Scientific Education and Research Network and among the top 2.5% of scientists by ResearchGate. His acheivements include: (i) the disintegrated melt deposition technique and (ii) the hybrid microwave sintering technique, an energy-efficient solid-state processing method to synthesize alloys or micro-/nano-composites. He has published over 565 peer-reviewed journal papers and owns two US patents and one trade secret. His current h-index is 66, his RG index is 47.4, and he has received more than 16,000 citations. He has also co-authored six books, published by John Wiley, Springer and MRF USA. He is editor-in-chief or editor for twelve international peer-reviewed journals. A multiple award winner, he actively collaborates with researchers in Japan, France, Saudi Arabia, Qatar, China, the USA and India as a visiting researcher, professor and chair professor.

Editorial

Metal Matrix Composites—The Way Forward

Manoj Gupta

Department of Mechanical Engineering, National University of Singapore, 9 Engineering Drive 1, Singapore 117576, Singapore; mpegm@nus.edu.sg

Received: 14 March 2020; Accepted: 12 April 2020; Published: 25 April 2020

The magic of the unification of different categories of materials to develop superior materials (composites) with improved functionality was recognized way back by our forefathers and was utilized very well by mother nature to support the dynamic functionality and requirements of both static and moving living organisms. Metal matrix composites (MMCs) represent one such unification type where the base material is either a metal or metallic alloy and the reinforcement is typically a metal or a ceramic. The development and application of MMCs critically depend on a number of factors. The most important of them are listed below [1–3]:

a.   Choice of primary and secondary processing techniques to synthesize MMCs;
b.   Choice of metallic matrix and reinforcement. They should be compatible to each other;
c.   Factors related to the reinforcement selection such as their type, shape, morphology, content and length scale;
d.   Matrix-reinforcement interfacial integrity;
e.   Heat treatment.

With the support of over 50 years of research in the area of MMCs, many industries such as transportation (land water and air), infrastructure, electrical, sports, machining, biomedical and space sectors are using MMCs increasingly in either the bulk or coating forms [4,5].

MMCs almost always exhibit a superior combination of certain properties such as strength, elastic modulus, hardness, tribological response and a controlled coefficient of thermal expansion. These properties in isolation or in certain combinations are useful for multiple engineering and biomedical applications, thus paving the way for materials scientists to continue with their work to develop even more superior MMCs. Having indicated this, it is of paramount importance to utilize fundamental metallurgical principles to judiciously unify matrix and reinforcement and to convert them into a final product.

Due to rapid ongoing technological advancements, service conditions are becoming more demanding, almost forcing the materials designers to make conscious efforts to develop materials including MMCs with improved properties. In order to highlight the current research efforts in the area of MMCs, *Applied Sciences*, has come out with a special thematic issue targeting MMCs exclusively. In the present issue, eight contributions from researchers are included that were accepted after rigorous peer review. The following metallic matrices were investigated:

a.   Magnesium-based matrix;
b.   Aluminum-based matrix;
c.   Iron-based matrix;
d.   Copper-based matrix;
e.   Sn-Ag-Cu-based matrix;
f.   Bronze-Ni-Mn-based matrix.

Note that these metallic matrices are utilized in a wide spectrum of engineering applications and are not restricted to any particular engineering sector. The basic inclination to use these matrices was to develop superior performance materials for:

a.   Weight-critical applications to minimize fuel consumption for healthy environments;
b.   Enhancing the performance of electronic solders;
c.   The development of high-performance tools;
d.   Thermal management in electronic applications.

The work presented in this Special Issue highlights the current trends and/or possible solutions that the industry is looking for, for increasing the reliability of their products further.

Overall, this book certainly provides a wonderful combination of high-quality research activities conducted across the world in the area of MMCs. This book is likely to be very useful for students and both young and experienced researchers in both academia and industry.

Last but not least, I would like to thank all the authors for their excellent contributions to this issue, to the reviewers for making very thoughtful comments and to the *Applied Sciences* editorial staff for publishing these articles at the earliest convenience.

Conflicts of Interest: The author declare no conflict of interest.

References

1. Ibrahim, I.A.; Mohamed, F.A.; Lavernia, E.J. Particulate reinforced metal matrix composites—A review. *J. Mater. Sci.* **1991**, *26*, 1137–1156. [CrossRef]
2. Lloyd, D.J. Particle reinforced aluminium and magnesium matrix composites. *Int. Mater. Rev.* **1994**, *39*, 1–23. [CrossRef]
3. Ceschini, L.; Dahle, A.; Gupta, M.; Jarfors, A.E.W.; Jayalakshmi, S.; Morri, A.; Rotundo, F.; Toschi, S.; Arvind Singh, R. *Aluminum and Magnesium Metal Matrix Nanocomposites*; Springer: Berlin, Germany, 2016; ISBN: 978-981-10-2680-5 (Print); ISBN: 978-981-10-2681-2 (Online).
4. Gupta, M.; Seetharaman, S. *Magnesium Based Nanocomposites for Cleaner Transport, Nanotechnology for Energy Sustainability*; Raj, B., Van de Voorde, M., Mahajan, Y., Eds.; Wiley-VCH: Weinheim, Germany, 2017.
5. Gupta, M.; Meenashisundaram, G.K. *Insight into Designing Biocompatible Magnesium Alloys and Composites*; SpringerBrief: Singapore, 2015.

Article

Effect of WC Nanoparticles on the Microstructure and Properties of WC-Bronze-Ni-Mn Based Diamond Composites

Siqi Li [1,2], **Zhe Han** [1,2], **Qingnan Meng** [1,2], **Xinzhe Zhao** [1,2], **Xin Cao** [1,2] **and Baochang Liu** [1,2,3,*]

1   College of Construction Engineering, Jilin University, Changchun 130026, China;
    siqi17@mails.jlu.edu.cn (S.L.); hanzhe17@mails.jlu.edu.cn (Z.H.); qingnanmeng@jlu.edu.cn (Q.M.);
    zhaoxz17@mails.jlu.edu.cn (X.Z.); caoxin16@mails.jlu.edu.cn (X.C.)
2   Key Laboratory of Drilling and Exploitation Technology in Complex Conditions of Ministry of Natural
    Resources, Changchun 130026, China
3   State Key Laboratory of Superhard Materials, Changchun 130021, China
*   Correspondence: liubc@jlu.edu.cn; Tel.: +86-431-8850-2891

Received: 24 July 2018; Accepted: 25 August 2018; Published: 1 September 2018

Abstract: Metal matrix-impregnated diamond composites are widely used for fabricating diamond tools. In order to meet the actual engineering challenges, researchers have made many efforts to seek effective methods to enhance the performance of conventional metal matrices. In this work, tungsten carbide (WC) nanoparticles were introduced into WC-Bronze-Ni-Mn matrix with and without diamond grits for improving the performance of conventional impregnated diamond composites. The influence of WC nanoparticles on the microstructure, densification, hardness, bending strength and wear resistance of matrix and diamond composites were investigated. The results showed that the bending strength of matrix increased up to approximately 20% upon nano-WC addition, while densification and hardness fluctuate slightly. The grinding ratio of diamond composites increased significantly by about 100% due to nano-WC addition. The strengthening mechanism was proposed according to experimental results. The diamond composites with 2.8 wt% nano-WC addition exhibited the best overall properties, thus having potential to apply to further diamond tools.

Keywords: WC nanoparticles; metal matrix; impregnated diamond composites; wear resistance

1. Introduction

Impregnated diamond composites produced by powder metallurgy are widely used for fabricating tools employed in cutting, drilling, milling and polishing applications [1–4]. The choice of the matrix materials which holds the diamond is essential to properties and service life of diamond tools [5,6]. WC-Bronze-Ni-Mn matrix, a composite mainly containing micron-WC, bronze (Cu85%-Sn6%-Zn6%-Pb3%) alloy, Ni and Mn, is widely used for diamond drilling tool manufacturing [7,8]. The micro-WC is used as framework material to enhance hardness and wear resistance, bronze is used as bonding phase, the element of Ni is used as strengthened phase because of its excellent wettability with diamond and Mn plays the role of antioxidant [9]. WC-Bronze-Ni-Mn matrix has high strength and adjustable properties that is suitable for different rock types, the diamond grits in tools can contact the rocks more easily and maintain an abrasive cutting surface between tools and rocks. The matrix properties and wear rate difference between matrix and diamond grits are important factors determining the performance of diamond tools [10–12]. Harsh and complex service conditions such as hydro-abrasive wear, impact stress and elevated temperature, require the development of new matrix materials with enhanced mechanical properties and wear resistance [13]. Different kinds of metals, such as Co, Fe and Ni, have been served as matrix materials for production of diamond tools [14–17]. However, the actual applicability of these matrices based diamond tools for different rocks is unsatisfactory because they wear out faster than diamond grits when processing hard and abrasive rocks. These metals also catalyze the reaction of diamond (sp3) to graphite (sp2) at elevated temperature, which reduces adhesion between diamond and matrix [18–20].

Development of metal matrix composites reinforced by nanoparticles is a promising way to meet the actual engineering requirement on mechanical and tribological properties. Metal matrix composites reinforced by nanoparticles have promising properties, which is more suitable for a large number of functional and structural applications than other metal matrix strengthening methods, such as solution strengthening, work hardening and precipitation strengthening [14]. Recently, some researchers investigated metal matrix reinforced with nano-particles (nanotubes), finding out that adding nanosized reinforcement improved hardness, bending strength and wear resistance of matrix [21–28]. Ceramic (Al_2O_3, ZrO_2, TiC, TiB_2, WC, etc.) nanoparticles can be used as reinforcing phase. But Al_2O_3, ZrO_2, TiC and TiB_2 nanoparticles have low wettability with bonding phase bronze, which decreases composites properties [7,13,22–26]. WC nanoparticles have excellent wettability with bronze and have good properties, namely high hardness, high wear resistance, good thermal and chemical stability. So it has been used in enhancing Ni, Fe and Co based matrix composites for cutting tools and modifying coatings for high wear resistance tools [13,22,29–31].

In this work, nano-WC was introduced into WC-Bronze-Ni-Mn based diamond composites for the first time. The effects of nano-WC addition on the microstructures, mechanical properties, and wear resistance of WC-Bronze-Ni-Mn based diamond composites were investigated and discussed. The aim is to seek optimal nano-WC addition concentration to meet the requirement of mechanical properties and wear resistance.

2. Materials and Methods

2.1. Sample Preparation

The composition of initial matrix in this work is given in Table 1, including WC (99.9% purity, Zhuzhou, China), bronze (99.9% purity, Beijing, China), Ni (99.9% purity, Beijing, China) and Mn (99.9% purity, Beijing, China).

Table 1. Composition of initial matrix.

Composition	WC	Bronze	Ni	Mn
Content (wt%)	55	35	5	5
Average particle size (μm)	10	50	75	60

Two series of samples (size: $38 \times 8 \times 5$ mm^3), matrix and impregnated diamond composites samples, were fabricated by power metallurgy methods of hot-pressing sintering. For purpose of obtaining uniform mixture, initial matrix powder was first mixed in a ball milling machine (Focucy, P400, Changsha, China) with WC balls for 24 h at a speed 120 rpm. Nano-WC (99.9% purity, average particle size of 80 nm, Qinhuangdao, China) with different mass percent concentration were then mixed with initial powder that was milled already using the same ball milling machine for 8 h at a speed 120 rpm. The resultant mixture with various concentration of nano-WC were sintering by hot-pressing in graphite moulds at 980 °C for 5 min. During the sintered process, the samples were forced by a uniaxial pressure of 35 MPa. The sintering apparatus was an intermediate frequency furnace (KGPS, Ezhou, China).

In impregnated diamond samples preparation process, initial matrix powder and different mass percent concentration nano-WC were mixed by the same way as matrix samples. The diamond grits (20 vol% concentration, synthetic, 270–325 μm, Changge, China) were added into the resultant mixture through a three-dimension mixer (JH2D-6, Zhengzhou, China) for 4 h. The diamond composites samples were prepared by hot-pressing sintering at the same parameters as matrix samples.

2.2. Characterization

The density of samples was measured by high precision density tester (Dahometer, DE-120M) via Archimedes method. The Rockwell hardness scale C (HRC) tests were carried out using a Rockwell hardness tester (Huayin HRS-150, Yantai, China). Three-point bending strength was tested by an electronic universal test machine (DDL 100, CIMACH, Chuangchun, China). The microstructure of composites were characterized by SEM (Hitachi S-4800, Tokyo, Japan) equipped with an energy dispersive spectrometer (EDS) and the acceleration voltage of EDS mapping experiments was 25 kV.

Grinding ratio tests were performed using a grinding ratio measurement apparatus as illustrated in Figure 1. The SiC grinding wheel with the dimension of 100 mm diameter and 20 mm thickness was applied as wear counterparts. The testing parameters were as follows: linear velocity 15 m/s, load 500 g, swing frequency 30 min^{-1} and grinding time > 100 s. The grinding ratio was calculated by the formula

$$Ra = \Delta Mg / \Delta Ms \tag{1}$$

where Ra is the sample grinding ratio; ΔMg is the weight loss of SiC grinding wheel; ΔMs is the weight loss of sample.

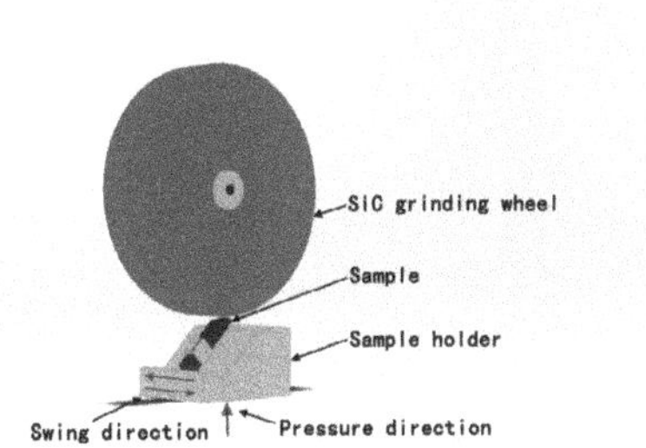

Figure 1. Schematic diagram of the grinding ratio test.

3. Results and Discussion

The designation, composition and mechanical properties of samples are summarized in Table 2.

Table 2. The designation, composition and mechanical properties of samples.

Samples	Composites	Relative Density (%)	Hardness (HRC)	Bending Strength (MPa)
S0	Matrix	98.4	42.2 ± 2.9	705.9 ± 35.4
S1	Matrix + 0.5 wt% nano-WC	97.0	36.9 ± 0.9	700.9 ± 30.3
S2	Matrix + 1.0 wt% nano-WC	97.0	36.7 ± 2.0	732.1 ± 16.6
S3	Matrix + 1.5 wt% nano-WC	97.8	39.2 ± 4.1	753.5 ± 31.0
S4	Matrix + 2.0 wt% nano-WC	97.6	36.2 ± 3.1	769.6 ± 29.9
S5	Matrix + 2.5 wt% nano-WC	98.5	40.7 ± 1.3	824.5 ± 28.4
S6	Matrix + 3.0 wt% nano-WC	98.3	38.8 ± 2.6	736.0 ± 24.5
SD0	Matrix + Diamond	95.8	-	391.0 ± 25.5
SD1	Matrix + 2.2 wt% nano-WC + Diamond	93.1	-	461.6 ± 32.0
SD2	Matrix + 2.5 wt% nano-WC + Diamond	94.8	-	398.8 ± 19.2
SD3	Matrix + 2.8 wt% nano-WC + Diamond	96.1	-	384.3 ± 7.4
SD4	Matrix + 3.1 wt% nano-WC + Diamond	95.3	-	398.0 ± 32.4

3.1. Microstructure

Sample S5 with 2.5 wt% nano-WC shows the optimum mechanical properties, so S5 and reference sample S0 are chose to be investigated. Figure 2 shows the fracture morphologies of matrix samples S0 and S5. In Figure 2a,b, the 0.5–6 μm matrix grains can be differentiated according to their shape: polygon-shaped grains consisting of micron-WC particles [7–9,29–31] and smooth round grains around, which are bonding phases bronze and Ni. As marked in Figure 2a, micro-WC particles are discerned from bronze phase grains by means of shape. The particle size of micro-WC in Figure 2 is from 0.5 to 6 μm that is not consistent with initial micro-WC particle size (10 μm). The ball milling process with WC balls is the reason of decrease in particle size. The ball milling machine (Focucy, P400, Changsha, China) is not a high energy ball mill and its main function is to mix powders evenly. So the wear amount of WC balls during ball milling process is extreme small and can not influence experimental results. Elements W, Cu, Ni and Mn show clear signals and uniform elemental distributions in EDS element mappings in Figure 3. The acceleration voltage used in EDS mapping experiments was 25 kV which correlates with a higher depth of analysis. So the element W is not clearly separated from other metals because the depth of analysis is larger than one layer particles.

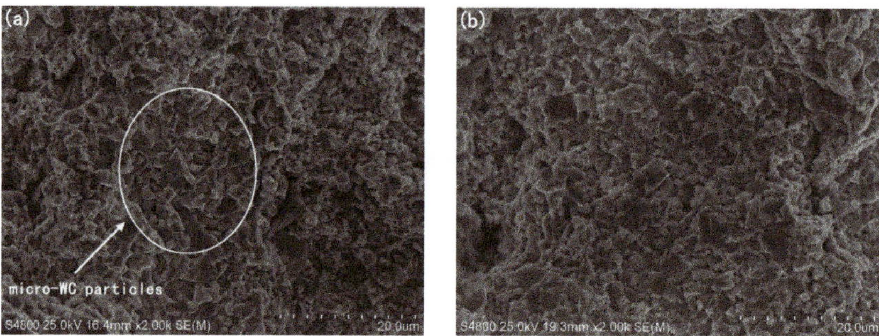

Figure 2. The fracture morphologies of samples: (**a**) S0; (**b**) S5.

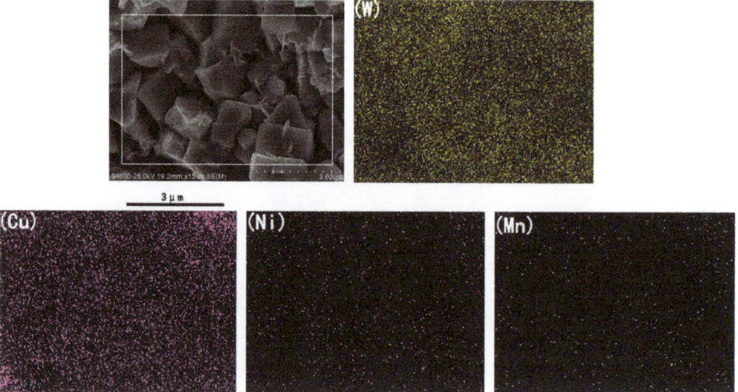

Figure 3. EDS element mappings of W, Cu and Ni for the fracture morphology of sample S5.

WC nanoparticles are found on fracture surface of sample S5 with 2.5 wt% nano-WC addition (Figure 4). The wettability between WC and bronze is excellent [9], so WC nanoparticles are wrapped by bronze phase during sintering process.

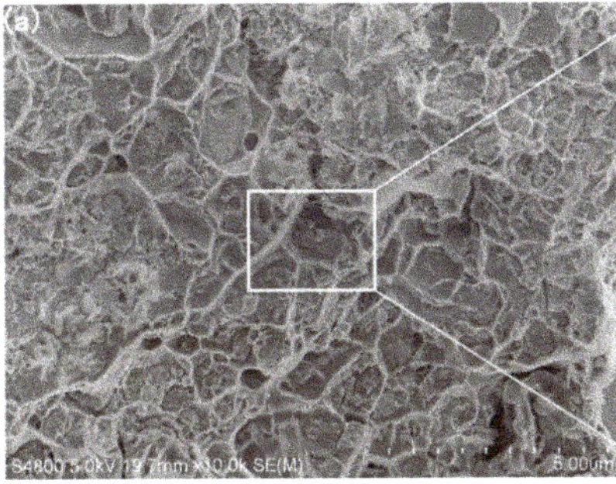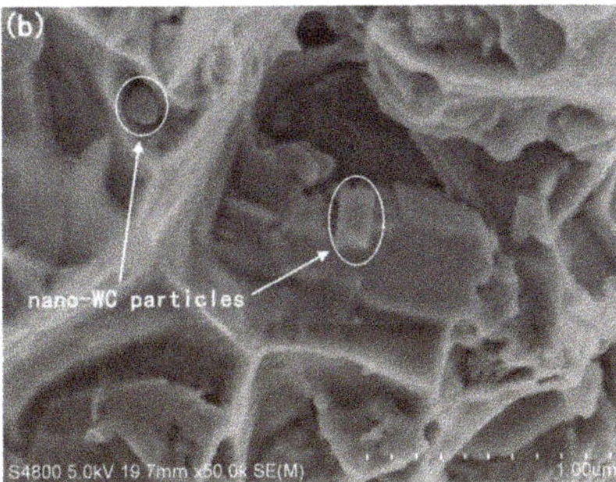

Figure 4. The fracture morphologies of sample: (**a**,**b**) S5.

SEM micrographs of impregnated diamond samples SD0 and SD1 are displayed in Figure 5. It shows that diamond grits are partially embedded in the matrix on fracture surfaces. As marked in Figure 5b,d, with the addition of 2.2 wt% nano-WC, the average width of the crack between diamond and matrix decreases from 4.1 μm to 3.0 μm. The smaller crack width means that the matrix holds diamond grits more firmly, which contributes positively to mechanical and tribological properties of impregnated diamond composites. The average width of the crack between diamond and matrix is shown in Table 3. The number of measured widths in the calculation of average is 15. With further increase of nano-WC addition, the influence on crack width is not significant. Some pores of SD1 found in Figure 5d have negative impact on relative density, which is consistent with the test results in Table 2. The pore existence states of sample SD2, SD3 and SD4 are similar to SD0.

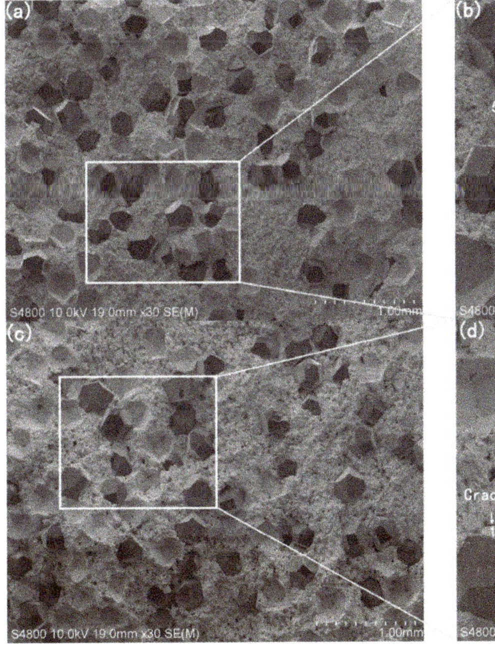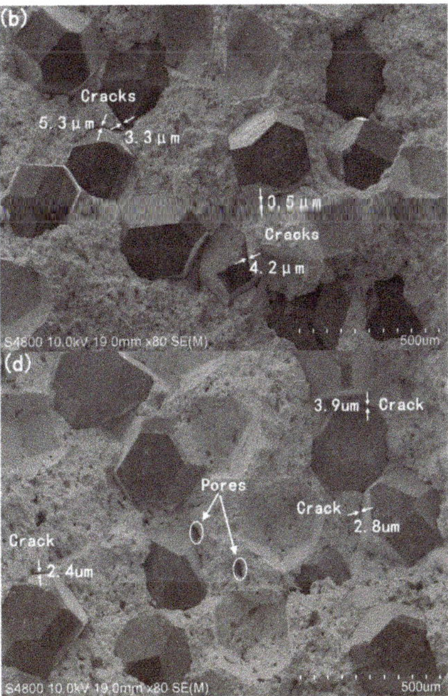

Figure 5. The fracture morphologies of samples: (**a**,**b**) SD0; (**c**,**d**) SD1.

Table 3. The average width of the crack between diamond and matrix of impregnated samples.

Samples	SD0	SD1	SD2	SD3	SD4
Width (μm)	4.1	3.0	3.8	4.0	4.2

3.2. Mechanical Properties

The mechanical properties of matrix samples are shown in Figure 6. With the increase of nano-WC particles addition, the relative density and hardness fluctuate and do not have significant changes (Figure 6a,b). It should be note that the introduction of nano-WC has little effect on relative density and hardness.

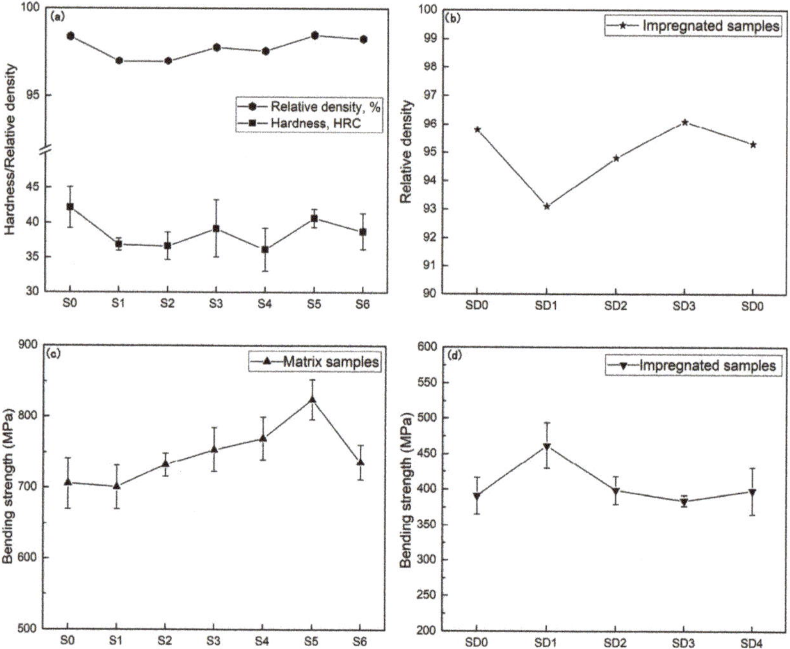

Figure 6. Mechanical properties of samples. (**a**) The relative density and HRC values of matrix samples; (**b**) The relative density of impregnated diamond samples; (**c**) The bending strength of matrix samples; (**d**) The bending strength of impregnated diamond samples.

Bending strength of matrix samples shows a trend of first increase and then decrease (Figure 6c), and sample S5 has the best value. The positive influence on bending strength of nano-WC addition can be associate with Orowan strengthening effect. The nano-WC particles pin the crossing dislocation and promote dislocations bowing around the particles under external load [32,33]. In addition, the mismatch in the coefficient of thermal expansion (CET) between WC ($6 \times 10^{-6}\,\mathrm{K}^{-1}$) and bronze ($\geq 18 \times 10^{-6}\,\mathrm{K}^{-1}$) needs further consideration [7,34]. During the cooling from processing temperature (980 °C), thermal stresses around the nano-WC particles lead to plastic deformation, especially in the interface area [33]. These stresses decrease quickly with increasing distance from the boundary, generating dislocation defects in the close vicinity of nano-WC. A large amount of nanoparticles is benefited to enhance dislocation density, resulting in an improvement of the deformation resistance. While concentration of nano-WC goes higher than 2.5 wt%, agglomeration phenomenon weakens the reinforcement effect.

As illustrated in Figure 6d, the bending strength of impregnated diamond sample SD1 shows the peak value and the value of SD2, SD3 and SD4 is close to SD0. Combined with fracture SEM observation (Figure 5) and average crack width (Table 3), the width of the crack decreases from 4.1 μm to 3.0 μm after adding 2.2 wt% nano-WC. This structural feature indicates matrix of SD1 has good diamond-holding capability, which is benefited to the stress transfer, enhancing the bending strength. Diamond grits act as weakening phase during bending strength test, so the holding strength at the interfaces between diamond grits and matrix is a factor influencing diamond composites bending strength and overall performance [3,5]. The interface structure observed between diamond grits and matrix of sample SD2, SD3 and SD4 is similar to SD0, this is accordance with bending strength test results.

3.3. Wear Resistance

Grinding ratio is a measurement index for evaluating the wear resistance and tribological performance of diamond composites [11]. The results for grinding ratio of impregnated diamond samples with different mass percent concentration nano-WC are summarized in bar chart (Figure 7). The grinding ratio increases remarkably after adding WC nanoparticles, proving that nano-WC plays a significant role in wear resistance of impregnated diamond composites. The grinding ratio increases firstly and then decreases with nano-WC content increases and grinding ratio of sample SD3 shows an about 100% increase in comparison with SD0. The wear resistance of impregnated diamond composites is higher than other WC-based, Fe-based, Co-based and Ni-based diamond composites [7,8,21–23].

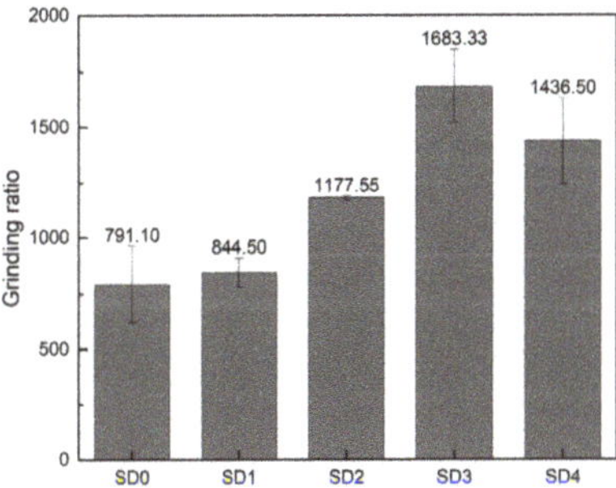

Figure 7. Results of grinding ratio test of impregnated diamond samples.

The diamond retention capacity largely determines the wear resistance of impregnated diamond composite [3]. As shown in Figure 4, WC nanoparticles are found on fracture surface, and these nanoparticles enhance the internal friction coefficient in the matrix as same as the friction coefficient between diamond grits and matrix [3,15]. Under complex and alternating cutting force, diamond grits show a tendency to rotate and then pull out of matrix easily. The increase of friction improves retentive capabilities of the matrix and reduces the rotating tendency, which enhances diamond retention in the matrix that is benefited to wear resistance of diamond composites. Hence, the grinding ratio of samples SD2, SD3 and SD4 with nano-WC addition is larger than reference one without nano-WC addition and SD3 with 2.8 wt% nano-WC has the best value. While concentration of nano-WC exceeds higher than 2.8 wt%, agglomeration phenomenon is harmful to the reinforcement effect.

The bending strength is also a factor affecting the wear resistance. The theoretical relationship between bending strength and wear resistance has been analyzed in the literatures [9–11]. A higher bending strength leads to a stronger support for diamond grits, meaning that diamond grits would not be lost permanently by pulling out when facing complex stresses conditions. But it can be found that the bending strength of SD1 is larger than SD0, SD2, SD3 and SD4 but the grinding ratio of SD1 with 2.2 wt% nano-WC is smaller than that of SD2, SD3 and SD4 and similar to SD0. This indicates that other factors weaken the wear resistance of SD1. As evidenced in Figure 5d or Figure 6b, the low relative density of SD1 means it has high porosity. The high porosity of SD1 implies the matrix wears faster than other samples, which causes diamond grits to pull out of matrix more easily. So the low relative density results in the decrease of grinding ratio of SD1.

4. Conclusions

The effect of WC nanoparticles on the microstructure and properties of WC-Bronze-Ni-Mn based impregnated diamond composites were investigated. Results showed that bending strength of matrix samples increased up to approximately 20% upon nano-WC addition, while densification and hardness fluctuate slightly. The grinding ratio of impregnated diamond composites increased significantly by about 100% due to nano-WC addition. The strengthening mechanism was proposed in detail. The related effects of bending strength, densification and diamond retention capabilities on wear resistance of diamond composites were revealed. WC-bronze impregnated diamond composite with 2.8 wt% nano-WC exhibited an optimal overall performance, thus having potential to apply to further diamond tools.

Author Contributions: B.L. and S.L. conceived and designed the experiments; S.L. and Z.H. performed the experiments; Q.M. and X.Z. analyzed the data; X.C. contributed analysis tools; S.L. wrote the paper.

Funding: This work is funded by the Science and Technology project of Jilin Province Education Department (JJKH20180087KJ) and the National Natural Science Foundation of China (41572357).

Conflicts of Interest: The authors declare no conflict of interest.

References

1. Tönshoff, H.K.; Hillmann, A.H.; Asche, J. Diamond tools in stone and civil engineering industry: Cutting principles, wear and applications. *Diam. Relat. Mater.* **2002**, *250*, 103–109. [CrossRef]
2. Boland, J.N.; Li, X.S. Microstructural Characterisation and Wear Behaviour of Diamond Composite Materials. *Materials* **2010**, *3*, 1390–1419. [CrossRef]
3. Zhao, X.; Duan, L. A Review of the Diamond Retention Capacity of Metal Bond Matrices. *Metals* **2018**, *8*, 307. [CrossRef]
4. Konstanty, J.S.; Baczek, E.; Romanski, A.; Tyrala, D. Wear-resistant iron-based Mn–Cu–Sn matrix for sintered diamond tools. *Powder Metall.* **2017**, *61*, 43–49. [CrossRef]
5. Artini, C.; Muolo, M.L.; Passerone, A. Diamond–metal interfaces in cutting tools: A review. *J. Mater. Sci.* **2011**, *47*, 3252–3264. [CrossRef]
6. Dai, H.; Wang, L.; Zhang, J.; Liu, Y.; Wang, Y.; Wang, L.; Wan, X. Iron based partially pre-alloyed powders as matrix materials for diamond tools. *Powder Metall.* **2015**, *58*, 83–86. [CrossRef]
7. Sun, Y.; Wu, H.; Li, M.; Meng, Q.; Gao, K.; Lu, X.; Liu, B. The Effect of ZrO_2 Nanoparticles on the Microstructure and Properties of Sintered WC-Bronze-Based Diamond Composites. *Materials* **2016**, *9*, 343. [CrossRef] [PubMed]
8. Li, M.; Sun, Y.H.; Dong, B.; Wu, H.D.; Gao, K. Study on effects of CNTs on the properties of WC-based impregnated diamond matrix composites. *Mater. Res. Innov.* **2015**, *19*, 59–63. [CrossRef]
9. Sun, Y.C.; Liu, Y.B. *Diamond Tool and Metallurgical Technology*; China Construction Materials Press: Beijing, China, 1999.
10. Di Ilio, A.; Togna, A. A theoretical wear model for diamond tools in stone cutting. *Int. J. Mach. Tools Manuf.* **2003**, *43*, 1171–1177. [CrossRef]
11. Wang, Z.; Zhang, Z.; Sun, Y.; Gao, K.; Liang, Y.; Li, X.; Ren, L. Wear behavior of bionic impregnated diamond bits. *Tribol. Int.* **2016**, *94*, 217–222. [CrossRef]
12. Ersoy, A.; Buyuksagic, S.; Atici, U. Wear characteristics of circular diamond saws in the cutting of different hard abrasive rocks. *Wear* **2005**, *258*, 1422–1436. [CrossRef]
13. Levashov, E.; Kurbatkina, V.; Alexandr, Z. Improved Mechanical and Tribological Properties of Metal-Matrix Composites Dispersion-Strengthened by Nanoparticles. *Materials* **2009**, *3*, 97–109. [CrossRef]

14. Casati, R.; Vedani, M. Metal Matrix Composites Reinforced by Nano-Particles—A Review. *Metals* **2014**, *4*, 65–83. [CrossRef]
15. Zhou, Y.M.; Zhang, F.L.; Wang, C.Y. Effect of Ni–Al SHS reaction on diamond grit for fabrication of diamond tool material. *Int. J. Refract. Met. Hard Mater.* **2010**, *28*, 416–423. [CrossRef]
16. Spriano, S.; Chen, Q.; Settineri, L.; Bugliosi, S. Low content and free cobalt matrixes for diamond tools. *Wear* **2005**, *259*, 1190–1196. [CrossRef]
17. Li, M.; Sun, Y.; Meng, Q.; Wu, H.; Gao, K.; Liu, B. Fabrication of Fe-Based Diamond Composites by Pressureless Infiltration. *Materials* **2016**, *9*, 1006. [CrossRef] [PubMed]
18. Uemura, M. An analysis of the catalysis of Fe, Ni or Co on the wear of diamonds. *Tribol. Int.* **2004**, *37*, 887–892. [CrossRef]
19. Narulkar, R.; Bukkapatnam, S.; Raff, L.M.; Komanduri, R. Graphitization as a precursor to wear of diamond in machining pure iron: A molecular dynamics investigation. *Comput. Mater. Sci.* **2009**, *45*, 358–366. [CrossRef]
20. Sidorenko, D.A.; Zaitsev, A.A.; Kirichenko, A.N.; Levashov, E.A.; Kurbatkina, V.V.; Loginov, P.A.; Rupasov, S.I.; Andreev, V.A. Interaction of diamond grains with nanosized alloying agents in metal–matrix composites as studied by Raman spectroscopy. *Diam. Relat. Mater.* **2013**, *38*, 59–62. [CrossRef]
21. Sidorenko, D.; Mishnaevsky, L.; Levashov, E.; Loginov, P.; Petrzhik, M. Carbon nanotube reinforced metal binder for diamond cutting tools. *Mater. Des.* **2015**, *83*, 536–544. [CrossRef]
22. Zaitsev, A.A.; Sidorenko, D.A.; Levashov, E.A.; Kurbatkina, V.V.; Rupasov, S.I.; Andreev, V.A.; Sevast'yanov, P.V. Development and application of the Cu-Ni-Fe-Sn-based dispersion-hardened bond for cutting tools of superhard materials. *J. Superhard Mater.* **2012**, *34*, 270–280. [CrossRef]
23. Loginov, P.; Mishnaevsky, L.; Levashov, E.; Petrzhik, M. Diamond and cBN hybrid and nanomodified cutting tools with enhanced performance: Development, testing and modelling. *Mater. Des.* **2015**, *88*, 310–319. [CrossRef]
24. Zaitsev, A.A.; Sidorenko, D.A.; Levashov, E.A.; Kurbatkina, V.V.; Andreev, V.A.; Rupasov, S.I.; Sevast'yanov, P.V. Diamond tools in metal bonds dispersion-strengthened with nanosized particles for cutting highly reinforced concrete. *J. Superhard Mater.* **2011**, *32*, 423–431. [CrossRef]
25. Loginov, P.A.; Sidorenko, D.A.; Levashov, E.A.; Petrzhik, M.I.; Bychkova, M.Y.; Mishnaevsky, L. Hybrid metallic nanocomposites for extra wear-resistant diamond machining tools. *Int. J. Refract. Met. Hard Mater.* **2018**, *71*, 36–44. [CrossRef]
26. Tjong, S.C. Novel Nanoparticle-Reinforced Metal Matrix Composites with Enhanced Mechanical Properties. *Adv. Eng. Mater.* **2007**, *9*, 639–652. [CrossRef]
27. Kamyar, S.; Mohammad, M.A.; Bhargav, P.; Minoo, N. Periodical patterning of a fully tailored nanocarbon on CNT for fabrication of thermoplastic composites. *Compos. A* **2018**, *107*, 304–314.
28. Karbalaei, A.M.; Rajabi, S.; Shirvanimoghaddam, K. Wear and friction behavior of nanosized Formula and Formula particle-reinforced casting A356 aluminum nanocomposites: A comparative study focusing on particle capture in matrix. *J. Compos. Mater.* **2015**, *49*, 3665–3681. [CrossRef]
29. Myalska, H.; Swadźba, R.; Rozmus, R.; Moskal, G.; Wiedermann, J.; Szymański, K. STEM analysis of WC-Co coatings modified by nano-sized TiC and nano-sized WC addition. *Surf. Coat. Technol.* **2017**, *318*, 279–287. [CrossRef]
30. Mauro, P.; Sasa, B. On the Properties of WC/SiC Multilayers. *Appl. Sci.* **2018**, *8*, 571. [CrossRef]
31. Kurlov, A.S.; Rempel, A.A. Effect of WC nanoparticle size on the sintering temperature, density, and microhardness of WC-8 wt % Co alloys. *Inorg. Mater.* **2009**, *45*, 380–385. [CrossRef]
32. Zhang, Z.; Chen, D. Consideration of Orowan strengthening effect in particulate-reinforced metal matrix nanocomposites: A model for predicting their yield strength. *Scr. Mater.* **2006**, *54*, 1321–1326. [CrossRef]
33. Zhang, Z.; Chen, D.L. Contribution of Orowan strengthening effect in particulate-reinforced metal matrix nanocomposites. *Mater. Sci. Eng. A* **2008**, *483–484*, 148–152. [CrossRef]
34. Vaidya, R.U.; Chawla, K.K. Thermal expansion of metal-matrix composites. *Compos. Sci. Technol.* **1994**, *50*, 13–22. [CrossRef]

 *applied sciences*

Article

Microstructures of Three In-Situ Reinforcements and the Effect on the Tensile Strengths of an Al-Si-Cu-Mg-Ni Alloy

Lusha Tian *, Yongchun Guo, Jianping Li *, Feng Xia, Minxian Liang, Hongbo Duan, Ping Wang and Jianli Wang

Shaanxi Province Engineering Research Centre for Aluminium/Magnesium Light Alloy and Composites, Materials and Chemical Engineering School, Xi'an Technological University, Xi'an 710021, China; yc-guo@163.com (Y.G.); xiafeng0811@163.com (F.X.); dlmxian@163.com (M.L.); 13679219266@163.com (H.D.); wping0922@163.com (P.W.); wjl810325@163.com (J.W.)
* Correspondence: tianlusha51@163.com (L.T.); lijianping@xatu.edu.cn (J.L.)

Received: 19 July 2018; Accepted: 27 August 2018; Published: 1 September 2018

Abstract: In the present paper, the microstructures of three kinds of in-situ reinforcements Al-Ti-C, Al-Ti-B, and Al-Ti-B-C-Ce were deeply investigated using a combination of scanning electron microscopy, X-ray diffraction spectroscopy, and transmission electron microscopy. The effect of in-situ reinforcements on the room temperature and elevated temperature (350 °C) tensile strengths of Al-13Si-4Cu-1Mg-2Ni alloy were analyzed. It is found that doping with trace amounts of B and Ce, the size of the Al_3Ti phase in the in-situ reinforced alloy changed from 80 μm (un-reinforced) to about 10 μm, with the simultaneous formation of the AlTiCe phase. The Al-Ti-B-C-Ce reinforcement which is rapid solidified, was more effective and superior to enhance the tensile strengths of the Al-13Si-4Cu-1Mg-2Ni alloy, both at room and high temperatures than those of addition other reinforcements. The room temperature (RT) strength increased by 19.0%, and the 350 °C-strength increased by 18.4%.

Keywords: Al-Ti-C; Al-Ti-B; Al-Ti-B-C-Ce; in-situ reinforcement; Al-Si-Cu-Mg-Ni alloy; microstructure; tensile strength

---

1. Introduction

In last two decades, in situ metal matrix composites, as a branch of discontinuous reinforced metal matrix composites, have been attracting many researchers interest and have made notable achievements [1]. The word "in situ composite" comes from "in situ crystallization". The concept of in-situ composite was proposed by Soviet scientist Merzhanov et al., where they synthesized TiB_2 reinforced Cu based functional gradient materials by Self-Propagating High-temperature Synthesis in 1987 [2–4]. High quality in situ aluminum-matrix and titanium-matrix composites have been developed for various applications in vehicles, aerospace, and the machinery industries [5,6]. However, the research on in situ metal matrix composites and its current research investment, are far from meeting the urgent needs of the high-tech development of modern industry and the military [7,8]. Therefore, it is of great importance to conduct more in-depth research on in situ metal matrix composites, for applications in leading-edge manufacturing and innovation areas, such as alloy material design theory, amorphous nano-crystalline matrix inoculation and refinement mechanism, and ceramic nano-particulate reinforcement. The goal is to make major breakthroughs rapidly in nano-fabrication technology and basic theoretical research in new types of metal matrix composite materials, so as to contribute to the transformation and upgrade of the traditional metal material industry [9–11].

The advantages of aluminum-based composite materials include high specific strength, excellent corrosion resistance, and outstanding electrical conductivity [12]. The primary purpose of adding reinforcement to the aluminum matrix, is to compensate for the low strength, inferior high-temperature performance, and poor wear resistance of aluminum. For aluminum matrix composites, the most effective particulate reinforcements are mostly ceramic particles. These particles can be divided into two types, one is directly formed in the matrix and the other is added after artificial synthesis. In addition to the high specific strength and specific modulus, these reinforcements have superior high-temperature performance; they are therefore widely used in the production of composite materials [13,14].

Table 1 shows the property parameter of the compound, which can be used as a particle enhancer. In recent decades, studies on TiC, TiB_2, ZrB_2, AlN or its biphasic particle composite reinforced aluminum matrix composites have been increasing. TiC particles, as the reinforced phase of in-situ aluminum matrix composites, show many excellent properties: high hardness, high melting point, high elastic modulus, low thermal conductivity, and good wettability and thermodynamic stability in molten aluminum [15]. TiB_2 has been intensively investigated in modern engineering materials, due to its high melting point, superior hardness, and excellent corrosion resistance, as well as good thermal and electrical conductivity. At present, TiB_2 is widely applied in the field of high-temperature electrodes, armor, cutting tools, and other structural materials [16].

Table 1. Properties of particle—enhancing compounds [17–19].

Particle	Density/(g·cm^{-3})	Melting Point/°C	Coefficient of Thermal Expansion/ (10^{-6} °C^{-1})	Thermal Conductivity	Elasticity Modulus/(GPa)
TiC	4.93	3160	7.74	24.28	450
TiB_2	4.25	2980	4.6–8.01	24.4–26	530
TiB	4.5	2473	8.6	-	550
SiC	3.19	2970	4.3	-	430
VC	5.36	2810	4.2	24.7	430
WC	15.55	2720	3.84	29.31	810
Al_2O_3	3.9	2050	8.6	28.89	420
Al_3Ti	3.3	1623	-	-	220

TiC and TiB_2 are usually synthesized in situ by Al-Ti-C and Al-Ti-B systems. At present, the most commonly used reinforcements in the industry are Al-Ti-C and Al-Ti-B reinforcements [20,21]. Table 2 shows the characteristics of various reinforcements [22–26]. Researches have been fully developed and widely used in industrial production. However, Al-Ti-C as well Al-Ti-B, have some disadvantages in the application process, where Al-Ti-B contains TiB_2 which tends to accumulate in molten aluminum, decreasing the refining effect. In addition, Al_3Ti particulates contained in Al-Ti-C and Al-Ti-B refinements have low refinement efficiency and degrade the matrix properties because of its needle-like appearance. New Al-Ti-B-C-Ce reinforcement, enhance and inherited the merits of the traditional reinforced body, have the effect of refining grain size, and improved its shortcoming to further improve the refining effect, and thus have a great advantage in improving the materials mechanical performance [27].

Table 3 shows the tensile properties of different alloys in the reference literature, at room temperature.

Al-13Si-4Cu-1Mg-2Ni alloy, is a widely used piston aluminum alloy through the enhanced in-situ method which further improves its mechanical properties. This is of great significance in meeting the research and development needs of high performance engines, such as high-power density diesel engines. In this paper, the strengthening effect of these three kinds of enhancers, and the addition to the piston alloy by in-situ generation, is studied to provide guidance for the development of a new piston alloy.

Table 2. Characteristics of reinforcements.

Reinforcement Types	Advantages	Disadvantages
Al-Ti-B	1. High refinement efficiency that is 12 times better than that of Al-Ti 2. $TiAl_3$ and TiB_2 are heterogeneous crystal cores of $\alpha(Al)$	1. Accumulation of TiB_2 can impair the quality 2. "Poisoning" by Zr, Cr, and Mn can lead to failure
Al-Ti-C	1. Non-contaminating addition of a refinement agent 2. "Immune" to Zr, Cr, and Mn 3. The grain size of TiC even smaller than that of TiB_2	1. Poor attenuation 2. Refinement effect worse than that of Al-Ti-B
Al-Ti-B-C-Ce	1. They have all the above mentioned advantages 2. Particle size is refined by adding rare earth to prevent aggregation and precipitation of $TiAl_3$ and TiB_2	1. Research not yet well developed

Table 3. The tensile properties of different in-situ reinforced aluminum matrix composites at room temperature.

Materials	Enhanced Phase Content/vol. %	Tensile Strength/MPa	Elongationδ/%	Reference
TiC/Al	15	610	4.9	[28]
TiC/Al-4.5Cu	20	540	18	[29]
TiB_2/Al	16	349	3.8	[30]
TiB_2/Al	5	124	9.2	[31]
TiB_2/A356	2.5	290	10	[32]
TiB_2/A356	5	302	9	[32]
TiB_2/A356	7.5	317	8	[32]
TiB_2/A356	10	328	6	[32]
TiB_2/ZL109	-	258.7	7.5	[33]

2. Experimental

2.1. Intermediate Alloy Preparation

The powders needed to prepare the intermediate alloy were Ti, C, KBF_4, K_2TiF_6, and B_4C powder. The Al-13Si-5Cu-1Mg-2Ni matrix alloy was prepared using highly pure Al (99.95 wt. %), pure Mg (99.9 wt. %), pure Si (99.9 wt. %), Al-50 wt. % Cu, and Al-10 wt. % Ni master alloys. The matrix alloy preparation process also used Al-20 wt. % Ce, Al-P alterative, mischmetal rich in La and Ce, and Hexachloroethane. Table 4 are the powders used to prepare intermediate alloys.

Table 4. Raw materials for chemical reagents in the experiment.

Powders	Particle Size	Purity	Manufacturer
Al	200 mesh	$\geq$99.0%	Shanghai shanpu chemical Co., Ltd. (Shanghai, China)
Ti	200–300 mesh	$\geq$99.0%	Sinopharm chemical reagent Co., Ltd. (Shanghai, China)
C	2000 mesh	$\geq$99.9%	Tianjin Kemiou Chemical Reagent Co., Ltd. (Tianjin, China)
K_2TiF_6	/	$\geq$99.0%	Shanghai SSS Reagent Co., Ltd. (Shanghai, China)
KBF_4	/	$\geq$99.0%	Shanghai SSS Reagent Co., Ltd. (Shanghai, China)
B_4C	100–300 mesh	$\geq$99.0%	Nangong ruiteng alloy material Co., Ltd. (Nangong, China)

For Al-Ti-C intermediate alloy, Ti powder and C powder in a Ti:C = 4:1 mass ratio and 10 wt. % were used. Al were measured according to the reaction formula:

$$Ti + C \rightarrow TiC \tag{1}$$

$$Al + Ti \rightarrow Al_3Ti \tag{2}$$

$$Al + C \rightarrow Al_4C_3 \tag{3}$$

These powders were mixed uniformly in a mortar and poured into a stainless-steel briquette mold (Φ 35 mm $\times$ 10 mm). The powder was pressed to form a powder briquette. The briquette was dried and placed in molten pure aluminum liquid at 1000 °C, for an in-situ reaction to produce Al-3Ti-0.25C intermediate alloy.

For Al-Ti-B intermediate alloy, KBF_4 and K_2TiF_6 were blended according to the reaction formula and stirred directly into 850 °C pure molten aluminum. The in-situ reaction then occurred in molten aluminum to produce the Al-Ti-B intermediate alloy.

$$K_2TiF_6 + 2KBF_4 \rightarrow TiB_2 + 4KF + 5F_2 \tag{4}$$

$$2K_2TiF_6 + 2KBF_4 + 5Al \rightarrow TiAl_3 + TiB_2 + 6KF + 2AlF_3 + 4F_2 \tag{5}$$

Al-Ti intermediate alloy was prepared by melting in a Model WK-II non-consumable vacuum electric arc furnace, according to a specific composition. B_4C powder wrapped in aluminum foil at 800 °C, was placed in a melting crucible to prepare Al-Ti-B-C intermediate alloy. Al-Ti-B-C-Ce was then prepared by melting Al-Ti-B-C intermediate alloy and Al-20Ce alloy, in the WK-II non-consumable vacuum arc furnace. The prepared block of Al-Ti-B-C-Ce reinforcement was then placed in a Model LZK-12A vacuum quenching furnace, for rapid solidification.

2.2. Experimental Process

2.2.1. The Process of Matrix Alloy

In the experimental study, an alloy model of Al-13Si-4Cu-1Mg-2Ni (wt. %) was prepared in an intermediate frequency induction melting furnace. Firstly, pure aluminum ingot blocks were added into the graphite crucible and were melted, then the pieces of pure silicon Al-10 wt. %Ni and Al-50 wt. %Cu intermediate alloys were added into the melt, respectively. Furthermore, mischmetal, rich in elements La and Ce, together with Al-P alloy as the modifier agents, were added into the molten alloy. Hexachloroethane as the degassing agent was added at 720–740 °C, and held for 15 min. Then Mg block was added under the melt completely, to keep it from contacting oxygen and burning. The crucible was removed with a crucible clamp and the slags were removed with a stainless-steel spoon. After skimming off slag, it was poured into a metal mold to obtain Φ 20 mm $\times$ 100 mm cast rods.

2.2.2. The Process of Composite Alloys

When the matrix alloy was melted, three different reinforcements Al-Ti-C, Al-Ti-B, and Al-Ti-B-C-Ce (5 wt. %) briquette were added in an aluminum foil to the molten aluminum. Figure 1 shows the preparation technology of the composite alloy. After skimming off slag, it was poured into a metal mold to obtain Φ 20 mm $\times$ 100 mm cast rods.

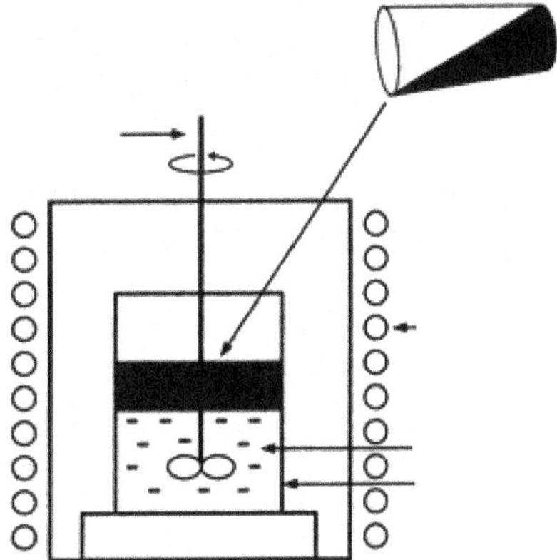

Figure 1. The preparation technology of composite alloy.

2.2.3. The Microstructure Analysis

The composite materials alloy when solidified, was analyzed using a Quanta 400 F (FEI, Hillsboro, OR, USA) thermal field emission scanning electron microscope equipped with an INCA energy dispersive spectrometer and JEM-2010 transmission electron microscope (JEOL, Tokyo, Japan) in detail. The phase compositions contained in the alloy prepared were examined using an XRD-6000 X-ray diffract meter (Shimadzu, Kyoto, Japan).

2.2.4. The Mechanical Tensile Tests

The tensile strengths of the cast Al-13Si-4Cu-1Mg-2Ni alloy at 25 °C and 350 °C, were tested with the Instron 1195 mechanical testing machine (Instron, Shanghai, China).

3. Results and Discussion

3.1. Metallurgical Analysis of Al-3Ti-0.25C Intermediate Alloy

3.1.1. X-ray Diffraction Analysis

Figure 2 shows the X-ray diffraction of the Al-Ti-C intermediate alloy. The XRD results show that there were Al, Al$_3$Ti, and TiC phases present in the Al-Ti-C intermediate alloy, prepared by the in-situ reaction.

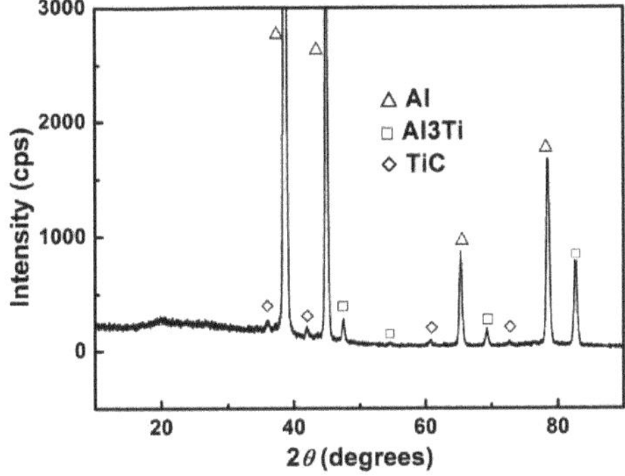

Figure 2. XRD analysis of Al-Ti-C intermediate alloy.

3.1.2. SEM Analysis

Figure 3 demonstrates a scan of the Al-Ti-C intermediate alloy, and energy spectral analysis of the different phases. The micrograph showed that the dimension of the long stripe phase was about 20–50 µm, and the maximum dimension of the small particle phase was approximately 1 µm. Figure 3a shows the energy spectrum of the long rod phase. The atomic ratio of Al:Ti was 3:1, and the long rod phase is Al_3Ti. Figure 3b shows the energy spectrum of the small particle phase. Herein, the grayish white near spherical nano-phase is TiC (Ti:C = 1:1). At 1200 °C, Ti and graphite react violently in the alloy melt, and a large amount of TiC crystals are formed in the early phase of the reaction. The spherical clusters mainly comprise Ti and C, with Ti:C = 49.08:50.92, which is close to the Ti:C ratio in the TiC compound. From the XRD results, it is believed that the spherical clusters are TiC phases. The reasons for the clustering may be that, in small particles, the large specific surface area leads to a high specific surface energy and they cluster to reduce the specific surface area and energy. In the meantime, calculations made using Image Pro showed that the volume fraction of Ti phase in the alloy was about 10.5%. Figure 3c shows the SEM image of the extracted Al-Ti-C intermediate alloy. Since the matrix had been dissolved and the other large phases were also removed by sieving, only TiC remained. The figure shows that TiC was mostly irregularly granular. Since the specimen was adsorbed on the copper mesh after extraction, the energy spectrum showed a strong uncalibrated Cu peak.

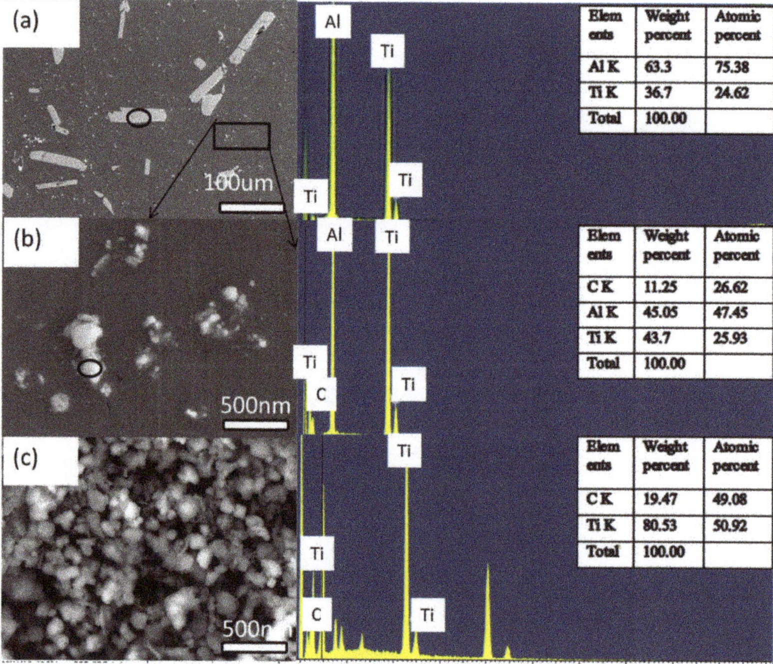

Elements	Weight percent	Atomic percent
Al K	63.3	75.38
Ti K	36.7	24.62
Total	100.00	

Elements	Weight percent	Atomic percent
C K	11.25	26.62
Al K	45.05	47.45
Ti K	43.7	25.93
Total	100.00	

Elements	Weight percent	Atomic percent
C K	19.47	49.08
Ti K	80.53	50.92
Total	100.00	

Figure 3. SEM micrograph and energy spectrum: (**a**) rod-like Al_3Ti phase; (**b**) grain-like TiC phase; (**c**) extracted TiC.

3.1.3. TEM Analysis

TiC is a face-centered cubic crystal; its close packed plane and close packed direction are {111} and [110], respectively. Since the (111) plane is a regular hexagonal base plane, the growth of a complete hexagonal TiC crystal is along the [111] direction, i.e., a coarse hexagonal TiC crystal grain is formed by hexagonal monolayers in a layered growth pattern along the [111] crystallographic direction. The formation of such a thick hexagonal TiC crystal is shown schematically in Figure 4.

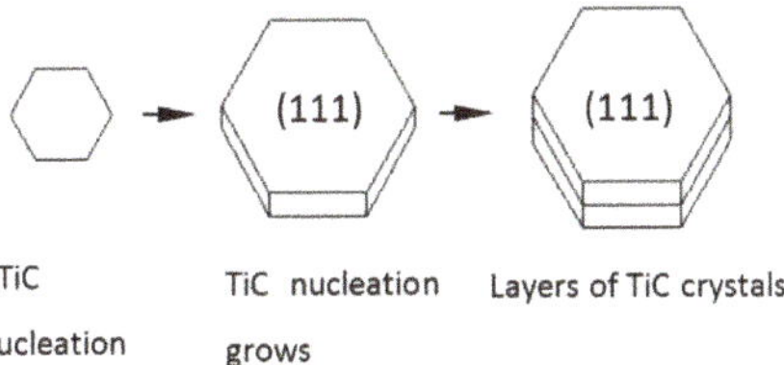

Figure 4. Schematic of the growth process of the hexagonal TiC crystal layers.

Figure 5 shows that the nano phases are distributed mainly on the grain boundaries. The size of the irregular hexagonal phase TiC was about 200 nm. At the same time, the diffraction pattern showed that the irregular hexagons mainly grew in the [011] and [21-1] directions.

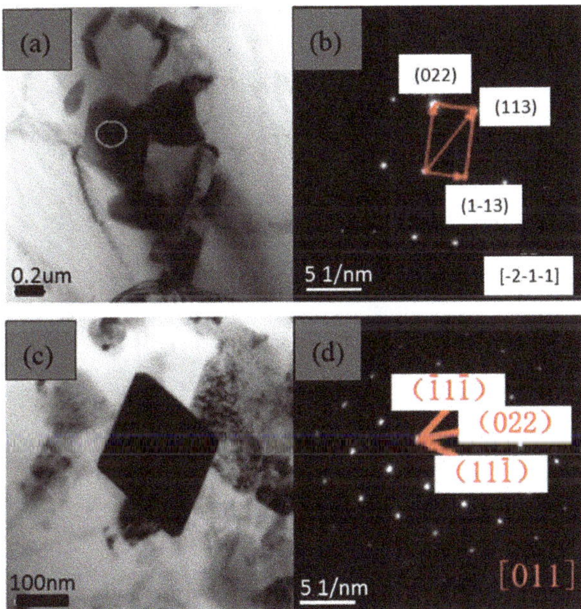

Figure 5. Nano-phase distribution and identification of diffraction spots. (**a**) TEM images of the nano phase; (**b**) diffraction spots of nano phase; (**c**) TEM images of the nano phase; (**d**) diffraction spots of nano phase.

3.2. Al-5Ti-B Intermediate Alloy Phase Analysis

3.2.1. X-ray Diffraction Analysis

Figure 6 shows the XRD results of Al-Ti-B master alloy. The XRD results showed that there were Al, Al_3Ti, and TiB_2 phases in the Al-Ti-B intermediate alloy, produced by the in-situ reaction.

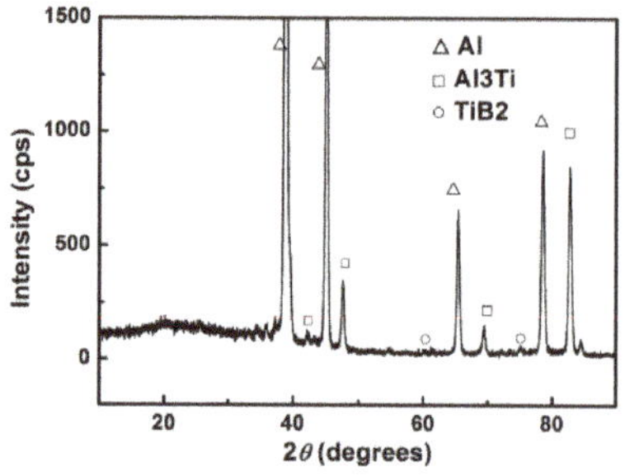

Figure 6. XRD spectrum of Al-Ti-B intermediate alloy.

3.2.2. SEM Analysis

Figure 7 shows the scan image of the Al-Ti-B intermediate alloy. It can be seen that there were two phases that were different in morphology and size, one of which was in the shape of a rod and the other was in the form of a grain. Figure 8 shows the scan image and energy spectrum of different phases. The energy spectrum showed that the atomic ratio of Al to Ti in the rod-shaped phase was 3:1. Further analysis showed that the long rod phase was Al_3Ti and the small grain was TiB_2. Furthermore, the extracted photos suggested that the granular TiB_2 was an irregular polygon with a constant thickness. The volume fraction of the Ti phase in the Al-Ti-B alloy was about 16.5%. Moreover, the Al_3Ti particles in Al-Ti-B alloy were smaller than the Al_3Ti particles in Al-Ti-C.

Figure 7. SEM image of the Al-Ti-B alloy. (**a**) Low magnification SEM; (**b**) High magnification SEM.

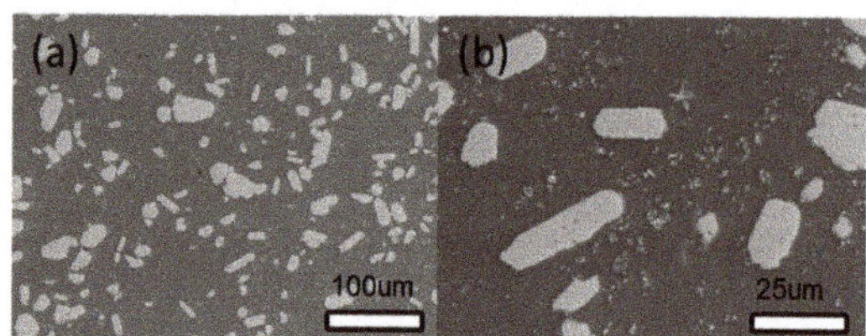

Elements	Weight percent	Atomic percent
Al K	63.3	75.31
Ti K	36.7	24.68
Total	100.00	

Elements	Weight percent	Atomic percent
B K	6.69	15.66
Al K	85.41	80.16
Ti K	7.9	4.18
Total	100.00	

Elements	Weight percent	Atomic percent
B K	33.27	68.84
Ti K	66.73	31.16
Total	100.00	

Figure 8. SEM image and energy spectrum of different phases: (**a**) rod-shaped Al_3Ti; (**b**) grain-shaped TiB_2; (**c**) extracted TiB_2.

3.2.3. TEM Analysis

Figure 9 shows the TEM and diffraction patterns of Al-Ti-B intermediate alloy. It showed that the nano-phase resides at the grain boundary and that the size of TiB$_2$ was about 200 nm. Moreover, there was a certain phase relationship between the Al (111) direction and the TiB$_2$ (100) direction.

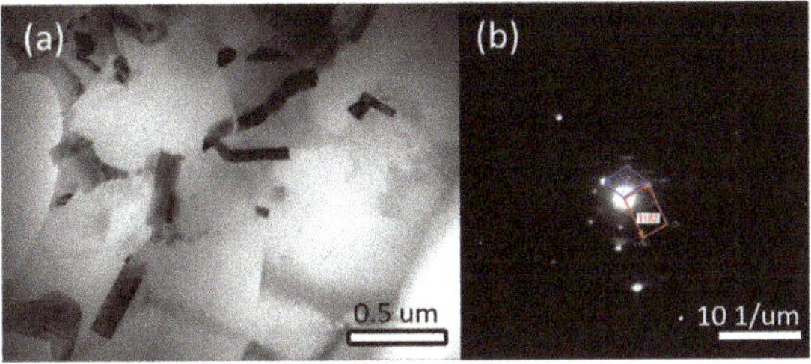

Figure 9. The TEM and diffraction patterns of Al-Ti-B intermediate alloy. (**a**) Low magnification SEM, (**b**) High magnification SEM.

3.3. *Al-5Ti-B-C-Ce Phase Analysis*

3.3.1. XRD Analysis

Figure 10 exhibits the X-ray diffraction pattern of Al-5Ti-B-C-Ce alloy. It indicated that the Al-Ti-B-C-Ce alloy included Al, Al$_3$Ti, TiB$_2$, and TiC phases. No Ce-containing phase was detected in the X-ray diffraction of Al-5Ti-B-C-Ce, due to the relatively low content of element Ce, which was only confirmed by energy spectrum analysis.

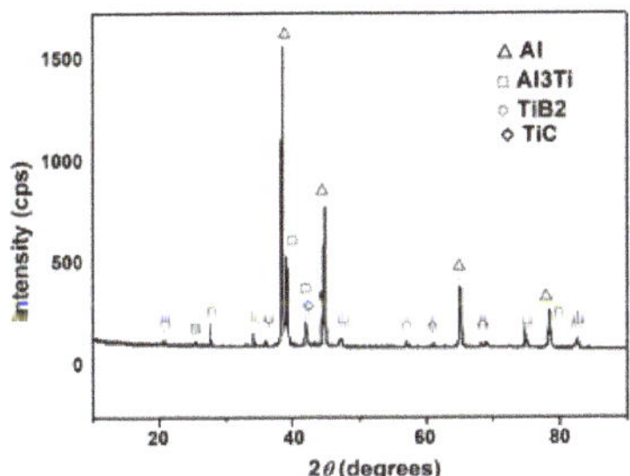

Figure 10. XRD spectrum of Al-Ti-B-C-Ce intermediate alloy.

3.3.2. SEM Analysis

Figure 11 shows the SEM image of the Al-5Ti-BC-Ce alloy specimen and EDX analysis. Figure 11a,b suggests that the nano-reinforcing particulates are distributed in the form of clusters. The particulates are small but numerous. The solidification process of the aluminum melt can provide more nucleation particles, which can be used as an effective intra-crystalline reinforcement to improve the performance of the alloy. The dark regions of the image are the aluminum substrates, and the bright colors refer to the ceramic particles. In Figure 11c1, the larger particles are found to be Al$_3$Ti from energy spectrum analysis. Figure 11c2 shows that trace amounts of rare earth element Ce can inhibit the rod-shaped Al$_3$Ti phase (80 μm is reduced to about 10 μm) and generate the Ti$_2$Al$_{20}$Ce phase. The energy spectrum of Figure 11d1, demonstrates that elements B and Ti are high in weight percentage, indicating that the larger particles in Figure 11d are TiB$_2$. Figure 11d2 shows that the weight percentage of elements C and Ti are high, indicating that the gray spot particles in Figure 11d are TiC. Figure 11e is the photomicrographs of the same extraction, showing a hexagonal prismatic phase, with numerous fine granular compounds attached around the hexagonal prism. The energy spectral analysis showed that it is composed of elements Ti, B, Al, and Cu, and has a high content of elements B, Al, and Cu in the spectrum, which are generated by the matrix and the net; therefore, the hexagonal prism is TiB_2. The energy spectrum analysis of the particles attached to the surface of the hexagonal prism, showed that it is composed of elements Ti, C, and B. The generation of B was mainly due to the small phase generated by the hexagonal prism. Therefore, the attached particles were TiC compounds. The energy spectrum analysis of the particles near the hexagonal prism, revealed that the particles were composed of elements C, Ti, and B, and the C content was high; so these fine particles were TiC phases. The results of scanning electron micrographs and energy spectra provided initial confirmation that TiB_2, TiC, and Al_3Ti particles were contained in the Al-5Ti-B-C-Ce alloy. The scan images also revealed the morphology of TiB_2 and TiC. Compared to Al-5Ti-C alone, the Al-5Ti-B and Al-5Ti-B-C-Ce phases were smaller and more uniform.

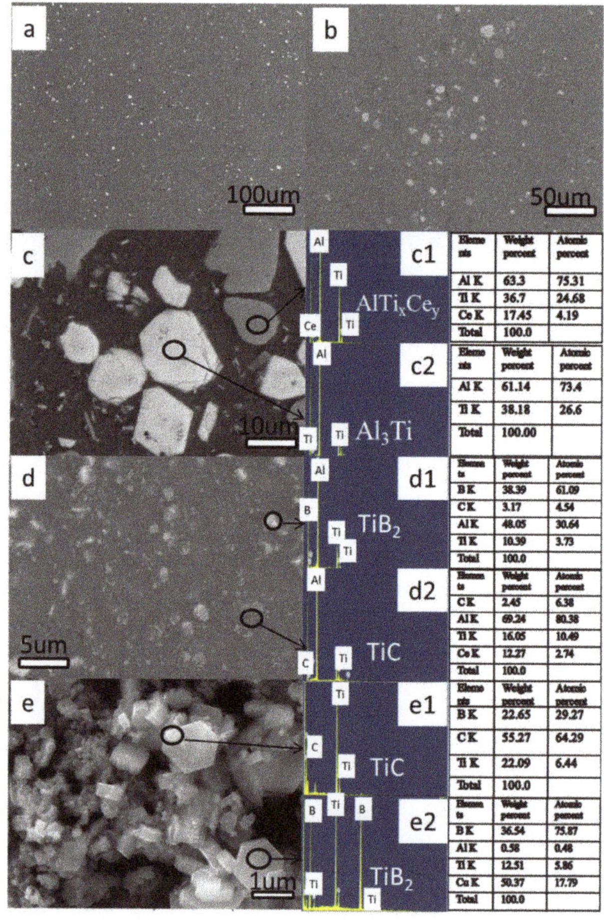

c1 — EDS of AlTi$_x$Ce$_y$

Elements	Weight percent	Atomic percent
Al K	63.3	75.31
Ti K	36.7	24.68
Ce K	17.45	4.19
Total	100.0	

c2 — EDS of Al$_3$Ti

Elements	Weight percent	Atomic percent
Al K	61.14	73.4
Ti K	38.18	26.6
Total	100.00	

d1 — EDS of TiB$_2$

Elements	Weight percent	Atomic percent
B K	38.39	61.09
C K	3.17	4.54
Al K	48.05	30.64
Ti K	10.39	3.73
Total	100.0	

d2 — EDS of TiC

Elements	Weight percent	Atomic percent
C K	2.45	6.38
Al K	69.24	80.38
Ti K	16.05	10.49
Ce K	12.27	2.74
Total	100.0	

e1 — EDS of TiC

Elements	Weight percent	Atomic percent
B K	22.65	29.27
C K	55.27	64.29
Ti K	22.09	6.44
Total	100.0	

e2 — EDS of TiB$_2$

Elements	Weight percent	Atomic percent
B K	36.54	75.87
Al K	0.58	0.48
Ti K	12.51	5.86
Cu K	50.37	17.79
Total	100.0	

Figure 11. SEM images and EDX analysis of Al-5Ti-B-C-Ce alloy specimens SEM images and EDX analysis of Al-5Ti-B-C-Ce alloy specimens. (**a**) Low magnification SEM; (**b**) High magnification SEM; (**c**) nano phases; (**c1**) EDS of AlTi$_x$Ce$_y$; (**c2**) EDS of Al$_3$Ti; (**d**) nano phases; (**d1**) EDS of TiB$_2$; (**d2**) EDS of TiC; (**e**)nano phases; (**e1**) EDS of TiC; (**e2**) EDS of TiB$_2$.

In the early formation stage of the TiC phase, in the high-temperature in situ reaction in the molten alloy, the TiC structure itself contains many C vacancies [34]. This caused instability in the TiC phase structure and poor reinforcement efficiency. According to reports, TiC can degrade rapidly in 30 min. Element B is adjacent to C in the periodic table and has similar properties. In a melt, element B may diffuse and blend into the TiC lattice. This provides a prerequisite for doping with small- sized B atoms, and a way to control the structure and performance of the TiC reinforcement phase. The size of TiC is about 200 nm and the size of TiB_2 ranges from 200–900 nm.

3.4. Effects of Three Different Reinforcements on the Tensile Properties of the Al Alloy

Figure 12 is a bar graph showing the tensile strength of the Al-13Si-4Cu-1Mg-2Ni aluminum alloy cast piston, with three different reinforcements. The RT tensile strength increased by 13.8%, 8.6%, and 19.0%, respectively. The 350 °C tensile strength increased by 3.4%, 2.3%, and 18.4%, respectively. It revealed that all three reinforcements were effective, and that the addition of Al-5Ti-B-C-Ce reinforcement was more effective than the other two reinforcements, in increasing the room temperature and high-temperature tensile strength.

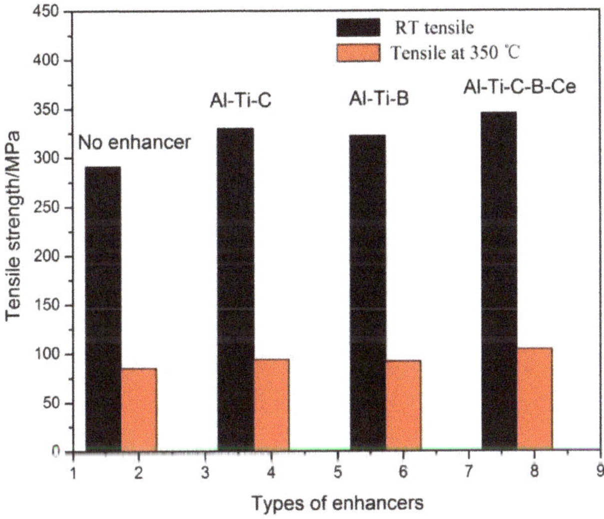

Figure 12. Tensile strengths of Al-13Si-4Cu-1Mg-2Ni alloy, with and without in situ reinforcements.

3.5. Strengthening Mechanism

In the process of solidification of the aluminum alloy, the dissolution and formation of intermetallic compounds will be formed, and the formed or added heteromorphic nuclear particles can be uniformly dispersed in the matrix alloy to promote the nucleation. By increasing the nucleation core of the non-uniform nucleation, it is precisely the principle of adding reinforcement to the matrix to refine the grain.

These lumpy Al_3Ti particles are block phase, which can cause cleavage phenomenon in the grain boundary and seriously affect the properties of the aluminum alloy [35]. To improve the performance of single Al-Ti-C and Al-Ti-B, the Al-Ti-B-C-Ce hybrid enhancement was prepared, which made TiC and TiB_2 play a double enhancement role. Simultaneously, the rare earths surface active elements can increase the wetting angle, improve the wettability of boride aluminum liquid, are easy to gather on the phase interface and grain interface adsorption, can fill in the surface of the grain defect, whilst making the grain continue to grow unhampered leading to the refinement of grain. Rare earth Ce, also has the effect of degassing (deoxygenation) [36,37]. In this way, TiC and TiB_2 are improved in the aluminum solution and the strength of the alloy is improved.

4. Conclusions

In this paper, three different kinds of reinforcements Al-Ti-C, Al-Ti-B, and Al-Ti-B-C-Ce were added to the alloy Al-13Si-4Cu-1Mg-2Ni, to compare its strengthening effect. The three reinforcements were also analyzed by SEM and TEM. The results are summarized as follows:

1. After doping with trace amounts of elements B and Ce, the size of reinforcement Al_3Ti phase changed from 80 μm (un-reinforced) to about 10 μm, with the simultaneous formation of $Ti_2Al_{20}Ce$ phase.

2. Since doping with element B destroys the equilibrium growth conditions for TiC phase, small nano phases of TiC attached to the surface and vicinity of hexagonal TiB_2. Compared with alloy without element B doping, the hexagonal TiB_2 phase turns out to be a growth carrier for TiC nano-phase, and the distribution of TiC nano-phase is more uniform.

3. All three in situ reinforcements were effective to the Al-13Si-4Cu-2Ni-1Mg alloy, and the addition of Al-5Ti-B-C-Ce reinforcement was more effective than the other two reinforcements in increasing the room temperature and high-temperature tensile strength. The RT strength increased by 19.0%, and the 350 °C-strength increased by 18.4%.

Author Contributions: L.T. wrote the main part of the manuscript and took part in the planning and execution of the experiments. Y.G. conceived and designed the study. J.L. participated in the coordination of the study and reviewed the manuscript. F.X. carried provided experimental alloys and analytical samples. M.L. analyzed part of experimental results. Transmission experiments were carried by H.D., P.W. and J.W. performed performance tests and made a final examination of the article.

Funding: The authors would like to thank financial support by the Creative Talents Promotion Plan—Technological Innovation Team (2017KCT-05), key research and development plan of Shaanxi province-Industrial project (2018GY-111) and key research and development plan of Shaanxi province project (2018ZDXM-GY-137).

Conflicts of Interest: The authors declare no conflicts of interest.

References

1. Clyne, T.W.; Withers, P.J. *An Introduction to Metal Matrix Composites: Applications*; Cambridge University Press: Cambridge, UK, 1993; p. 454.

2. Moore, J.J.; Feng, H.J. Combustion synthesis of advanced materials: Part I. Reaction parameters. *J. Prog. Mater Sci.* **1995**, *39*, 243–273. [CrossRef]

3. Merzhanov, A.G. Self-propagating high temperature synthesis and powder metallurgy: Unity of goals and competition of principles. *Adv. Powder Metall. Part Mater.* **1992**, *9*, 341–368.

4. Merzhanov, A.G. Academician. Processes of self-propagating high-temperature synthesis at new frontiers Academician. *Ind. Ceram.* **2009**, *29*, 196–199.

5. Nampoothiri, J.; Harini, R.S.; Nayak, S.K.; Raj, B.; Ravi, K.R. Post in-situ reaction ultrasonic treatment for generation of Al-4.4Cu/TiB_2 nanocomposite: A route to enhance the strength of metal matrix nanocomposite. *J. Alloys Compd.* **2016**, *683*, 370–378. [CrossRef]

6. Tjong, S.C.; Mai, Y.W. Processing-structure-property aspects of particulate and whisker-reinforced titanium matrix composites. *Compos. Sci. Technol.* **2008**, *68*, 583–601. [CrossRef]

7. Ramesh, C.S.; Keshavamurthy, R.; Channabasappa, B.H.; Pramod, S. Influence of heat treatment on slurry erosive wear resistance of Al6061 alloy. *Mater. Des.* **2009**, *30*, 3713–3722. [CrossRef]

8. Macke, A.; Schultz, B.F.; Rohatgi, P. Metal matrix composites offer the auto-motive industry an opportunity to reduce vehicle weight, improve performance. *Adv. Mater. Processes* **2012**, *170*, 19–23.

9. Yang, J.; Pan, L.; Gu, W.; Qiu, T.; Zhang, Y. Microstructure and mechanical properties of in situ synthesized (TiB_2 + TiC)/Ti_3SiC_2 composites. *Ceram. Int.* **2012**, *38*, 649–655. [CrossRef]

10. Choi, Y.; Rhee, S.W. Effect of carbon sources on the combustion synthesis of TiC. *J. Mater. Sci.* **1993**, *28*, 6669–6675. [CrossRef]

11. Kaftelen, H.; Unlu, N.; Goller, G. Comparative processing-structure–property studies of Al-Cu matrix composites reinforced with TiC particulates. *Compos. Part A Appl. Sci. Manuf.* **2011**, *42*, 812–824. [CrossRef]

12. Liu, Z.; Han, Q.; Li, J. Fabrication of in situ Al_3Ti/Al composites by using ultrasound assisted direct reaction between solid Ti powders and liquid Al. *Powder Technol.* **2013**, *247*, 55–59. [CrossRef]

13. Tang, P.; Li, W.F.; Wang, K.; Du, J.; Chen, X.Y.; Zhao, Y.J.; Li, W.Z. Effect of Al-Ti-C master alloy addition on microstructures and mechanical properties of cast eutectic Al-Si-Fe-Cu alloy. *Mater. Des.* **2017**, *115*, 147–157. [CrossRef]

14. Sadeghi, E.; Karimzadeh, F.; Abbasi, M.H. Thermodynamic analysis of Ti-Al-C intermetallics formation by mechanical alloying. *J. Alloys Compd.* **2013**, *576*, 317–323. [CrossRef]

15. Rai, R.N.; Datta, G.L.; Chakraborty, M.; Chattopadhyay, A.B. A study on the machinability behaviour of Al-TiC composite prepared by in situ technique. *Mater. Sci. Eng. A* **2006**, *428*, 34–40. [CrossRef]

16. Li, P.T.; Li, Y.G.; Wu, Y.Y.; Ma, G.L.; Liu, X.F. Distribution of TiB_2 particles and its effect on the mechanical properties of A390 alloy. *Mater. Sci. Eng. A* **2012**, *546*, 146–152. [CrossRef]

17. Liu, Z.D.; Tian, J.; Li, B.; Zhao, L.P. Microstrcuture and mechanical behaviors of in situ TiC particulates reinforced Ni matrix composites. *Mater. Sci. Eng. A* **2010**, *527*, 3898–3903. [CrossRef]

18. Wangjara, P. Characterization of Ti-6% Al-4%V/TiC Particulate Reinforced Metal Matrix Composites Consolidated by Sintering and Thermomechanical Processing. Ph.D. Thesis, Mcgill University, Montreal, QC, Canada, 1999.

19. Pearson, W.B. *A Handbook of Lattice Spacings and Structures of Metals and Alloys*; Pergamon Press: Oxford, UK, 1958.

20. Ramesh, C.S.; Pramod, S.; Keshavamurthy, R. A study on microstructure and mechanical properties of Al 6061—TiB_2 in-situ composites. *Mater. Sci. Eng. A* **2011**, *528*, 4125–4132. [CrossRef]

21. Wang, E.Z.; Gao, T.; Nie, J.F.; Liu, X.F. Grain refinement limit and mechanical properties of 6063 alloy inoculated by Al-Ti-C (B) master alloys. *J. Alloys Compd.* **2014**, *594*, 7–11. [CrossRef]

22. Zhao, H.L.; Song, Y.; Li, M.; Guan, S.K. Grain refining efficiency and microstructure of Al-Ti-C-RE master alloy. *J. Alloys Compd.* **2010**, *508*, 206–211. [CrossRef]

23. Tabrizi, S.G.; Sajjadi, S.A.; Babakhani, A.; Lu, W. Influence of spark plasma sintering and subsequent hot rolling on microstructure and flexural behavior of in-situ TiB and TiC reinforced Ti_6Al_4V composite. *Mater. Sci. Eng. A* **2015**, *624*, 271–278. [CrossRef]

24. El-Mahallawy, N.; Taha, M.A.; Jarfors, A.E.W.; Fredriksson, H. On the reaction between aluminium K_2TiF_6 and KBF_4. *J. Alloys Compd.* **1999**, *292*, 221–229. [CrossRef]

25. Xuan, Q.Q.; Shu, S.L.; Qiu, F.; Jin, S.B.; Jiang, Q.C. Different strain-rate dependent compressive properties and work-hardening capacities of 50 vol% TiC_x/Al and TiB_2/Al composites. *Mater. Sci. Eng. A.* **2012**, *538*, 335–339. [CrossRef]

26. Gao, Q.; Wu, S.S.; Lu, S.L.; Duan, X.C.; Zhong, Z.Y. Preparation of in-situ TiB_2 and Mg_2Si hybrid particulates reinforced Al-matrix composites. *J. Alloys Compd.* **2015**, *651*, 521–527. [CrossRef]

27. Nie, J.F.; Liu, X.F.; Ma, X.G. Influence of trace boron on the morphology of titanium carbide in an Al-Ti-C-B master alloy. *J. Alloys Compd.* **2010**, *491*, 113–117. [CrossRef]

28. Wu, Z.S. Preparation and Properties of TiC-Based Composites. Ph.D. Thesis, Jiangsu University, Nanjing, China, 2004.

29. Ma, M.Z.; Cai, D.Y.; Wei, T.H. Microstructure and mechanical properties of TiC_p/Al-4.5Cu-0.8Mg compsites by direct reaction synthesis. *J. Mater. Sci. Technol.* **2003**, *19*, 447–449.

30. Le, Y.K.; Zhang, Y.Y. Research status of particle reinforced aluminum matrix composites. *Dev. Appl.* **1997**, *5*, 33–37.

31. Tee, K.L.; Lu, L.; Lai, M.O. Synthesis of in situ Al-TiB_2 composites using stir cast route. *Compos. Struct.* **1999**, *47*, 589–593. [CrossRef]

32. Mandal, A.; Chakraborty, M.; Murty, B.S. Ageing behaviour of A356 alloy reinforced with in-situ formed TiB_2 particles. *Mater. Sci. Eng. A* **2008**, *489*, 220–226. [CrossRef]

33. Zhao, D.G.; Liu, X.F.; Bian, X.F. Preparation and mechanical properties of TiB_2 + SiC/ZL109. *Casting* **2004**, *53*, 97–100.

34. Song, W.; Wang, L.; Chen, X.; Chen, Y.; Li, Q. Effects of Al-Ti-C-Ce master alloy on As-cast microstructure and mechanical properties of 7010 aluminum alloy. *Spec. Cast. Nonferrous Alloy* **2011**, *4*, 39.

35. Cheng, L.H.; Sun, Y.; Sun, G.X. Effect of Al-5Ti-1B on the microstructure of near-eutectic Al-13%Si alloys modified with Sr. *J. Mater. Sci.* **2002**, *37*, 3489–3495.
36. Wang, K.; Cui, C.; Wang, Q.; Liu, S.; Gu, C. The microstructure and formation mechanism of core–shell-like TiAl$_3$/Ti$_2$Al$_{20}$Ce in melt-spun Al-Ti-B-Re grain refiner. *Mater. Lett.* **2012**, *85*, 153–156. [CrossRef]
37. Yin, J.B.; Li, Y.T.; Bi, J.M.; Pang, X.Z.; Li, L.J. Research on grain refinement of Al-2Ti-B-Ce and Al-5Ti-B-Ce master alloys on industrial pure aluminum. *Light Alloy Fabr. Technol.* **2015**, *1*, 5.

Article

Microstructure and Mechanical Properties of Magnesium Matrix Composites Interpenetrated by Different Reinforcement

Shuxu Wu [1], Shouren Wang [1,*], Daosheng Wen [1], Gaoqi Wang [1] and Yong Wang [2]

1   School of Mechanical Engineering, University of Jinan, Jinan 250022, China; Shuxu1994@gmail.com (S.W.); me_wends@ujn.edu.cn (D.W.); me_wanggq@ujn.edu.cn (G.W.)
2   School of Physics and Technology, University of Jinan, Jinan 250022, China; ss_wangy@ujn.edu.cn
*   Correspondence: me_wangsr@ujn.edu.cn

Received: 15 September 2018; Accepted: 19 October 2018; Published: 23 October 2018

Abstract: The present work discusses the microstructure and mechanical properties of the as-cast and as-extruded metal matrix composites interpenetrated by stainless steel (Fe–18Cr–9Ni), titanium alloy (Ti–6Al–4V), and aluminum alloy (Al–5Mg–3Zn) three-dimensional network reinforcement materials. The results show that the different reinforcement materials have different degrees of improvement on the microstructures and mechanical properties of the magnesium matrix composites. Among them, magnesium matrix composites interpenetrated by stainless steel reinforcement have maximum tensile strength, yield strength, and elongation, which are 355 MPa, 241 MPa, and 13%, respectively. Compared with the matrix, it increases by 47.9%, 60.7% and 85.7%, respectively. Moreover, compared with the as-cast state, the as-extruded sample has a relatively small grain size and a uniform size distribution. The grain size of the as-cast magnesium matrix composites is mainly concentrated at 200–300 µm, whereas the extruded state is mainly concentrated at 10–30 µm. The reason is that the coordination deformation of reinforcement and matrix, and the occurrence of dynamic recrystallization, cause grain refinement of magnesium matrix composite during the extrusion process, thereby improving its mechanical properties. Moreover, the improvement is attributed to the effect of the reinforcement itself, and the degree of grain refinement of the metal matrix composites.

Keywords: microstructure; mechanical properties; interpenetrating composites; magnesium alloy; metal reinforcement; metal matrix composites

---

1. Introduction

The metal matrix composites (MMCs) are a new type of structural material which is compounded by metal and alloy, as a matrix, through a certain process. The composite material has outstanding advantages, such as high specific modulus and specific strength, which is widely used in the fields of aviation, aerospace, and military protection [1,2]. In the area of the matrix, most metallic systems have been explored for use in metal matrix composites, including Al, Be, Ti, Mg, Fe, Ni, Co, and Ag, and, by far, the most mature technology is the preparation of aluminum matrix composites [3]. However, the traditional composite materials are usually reinforced by particles, fibers, and whiskers [4–6]. In recent years, with the requirement of the "structure–function" integration of composite materials and the development trend of increasing the volume fraction of the second phase particles, a new type of composite material has emerged, namely, the network interpenetrating structure of the reinforced the composite material. There is a new reinforcement method in which MMCs penetrate each other through three-dimensional network reinforcements, which have the mechanical characteristics of continuous enhancement, and the microstructure characteristics of intertwisting and interpenetrating with the matrix [7–9]. Since the matrix and reinforcement form a three-dimensional network structure in which the respective continuous and interpenetrating structures are formed in the space, the characteristics of each component constituting the network structure remain substantially unchanged, so its comprehensive reinforcement effect is generally better than the traditional reinforcement method, which makes it possible to obtain high-performance, multi-functional composite materials [10,11]. However, up to now, the metal matrix has been almost reinforced by ceramic, carbon nanotubes, or another brittle three-dimensional network reinforcement [12–14]. Due to the low toughness and shock resistance of brittle materials, these interpenetrating composites cannot undergo more machining deformation [15]. Particularly, in the case of fabricating structures having irregular cross-sectional shapes, their applications will be limited.

In order to solve the above problems, the further development direction of the network interpenetrating metal matrix composites may be on the choice of the reinforcement materials, which can not only improve the strength of the composites, but also improve the plasticity [16]. Therefore, when metal is selected as a reinforcement for the composites, the metal matrix composites interpenetrated by metal reinforcement (MIMC) will have good forgeability and workability, which can be processed and deformed by multiple machining processes, such as torsion straining (TS), warm rolling (WR), twin-roll casting (TRC), and reciprocal extrusion (RE), etc. [17–19]. Among them, the extrusion process of the metal matrix composites can significantly improve the combined quality between matrix and reinforcement [20,21]. However, few researchers have studied the microstructure and mechanical properties of MIMC, especially magnesium matrix composites [22,23]. At the same time, magnesium and its alloy have the advantages of being lightweight, and having good casting performance, good damping, and shock absorbability, but their mechanical strength is poor [24,25]. Therefore, this article proposes a network interpenetrating bimetallic magnesium matrix composite using three metal-made network skeletons as reinforcements. This bimetallic magnesium matrix composite can not only make up for the deficiency of magnesium alloy in terms of strength and hardness, but also has a better toughness and absorbing properties as a metal material than ceramic materials.

Therefore, the current work is mainly to manufacture the magnesium matrix composites interpenetrated by steel reinforcement, titanium reinforcement, and aluminum reinforcement and the subsequent extrusion process and, then, study the effects of different three-dimensional network reinforcement on the mechanical properties and microstructure of the magnesium matrix composites.

2. Materials and Methods

2.1. MIMC Composites Fabrication

Previously, our team used a three-dimensional network of Si_3N_4 ceramic structure and pressure-assisted and vacuum-driven permeation technique to reinforce magnesium matrix composites (Si_3N_4–Mg) [26]. The preparation process was as follows: a porous ceramic skeleton was prepared from high-purity β-Si_3N_4 powder (97% of Si_3N_4 and less than 100 μm in diameter, Shanghai Silicon Materials Plant, Shanghai, China). The porous reticulated polyurethane (PU) was selected as the original framework, and the porous prefabricated parts were prepared through the replication process. Figure 1a shows the porous network of Si_3N_4 ceramic skeleton after impregnation and sintering. Then, it was heated in a nitrogen-filled muffle furnace (SQFL-1700, Shanghai, China) until it reached the desired processing temperature of 750 °C, at which time, the liquid magnesium alloy was infiltrated into the preform skeleton by means of pressure-assisted and vacuum-driven permeation techniques. Figure 1b shows a Si_3N_4–Mg metal matrix composite, in which zone "1" is the Si_3N_4 three-dimensional network skeleton, and zone "2" is the Mg matrix. Figure 1c shows the schematic diagram of the infiltration device, and the mechanical properties of Si_3N_4–Mg matrix composite are shown in Table 1, which indicates that the ceramic network reinforcement significantly enhances the elastic modulus, rockwell hardness, and tensile strength of the matrix, while the elongation is decreased [27,28].

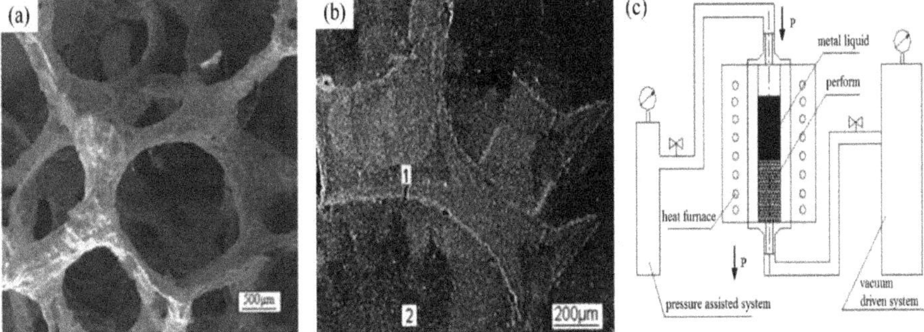

Figure 1. SEM micrographs image and technical diagram: (**a**) Si_3N_4 reticulated ceramic skeleton; (**b**) Si_3N_4–Mg composite; (**c**) the schematic diagram of infiltration equipment.

Table 1. The mechanical properties of Mg alloy and Si_3N_4–Mg composite.

Materials	Elastic Module (GPa)	Rockwell Hardness (HRB)	Elongation (%)	Tensile Strength (MPa)
Mg alloy	70	65	9.5	330
12Si$_3$N$_4$–Mg	110	71	3.2	345
25Si$_3$N$_4$–Mg	133	78	1.6	340

Recently, however, our team tried a new and simple process to fabricate the magnesium matrix composites. The network skeleton of the reinforcement designed by three-dimensional weaving technology was shown in Figure 2a, and the model of the metal matrix composite of the infiltrated reinforcement was shown in Figure 2b. The reinforcement skeleton materials were chosen as stainless steel (Fe–18Cr–9Ni), titanium alloy (Ti–6Al–4V), and aluminum alloy (Al–5Mg–3Zn), respectively, while magnesium matrix was selected as AZ31(Mg–3.31Al–1.05Zn). The skeleton was fabricated using the three-dimensional weaving techniques, and measured by Archimedes measurements (ASTM C373), with a reinforcement volume fraction of approximately 15%. Unlike the preparation of Si_3N_4–Mg matrix composites, the MIMC composites' fabrication processing was applied pressure infiltration technology, which is shown in Figure 2c. The entire experiment was carried out under an inert gas argon atmosphere. Firstly, in the inert atmosphere of a mixture of CO_2 and SF_6, the furnace temperature was set to 720 °C to melt the AZ31 alloy. Then, the molten magnesium alloy liquid was poured into an infiltration mold having a temperature of 650 °C and, finally, the pressure head started to be pressed. At last, the magnesium matrix composites interpenetrated by stainless steel reinforcement (MISC), the magnesium matrix composites interpenetrated by titanium alloy reinforcement (MITC), and the magnesium matrix composites interpenetrated by aluminum alloy reinforcement (MIAC), were fabricated successfully.

2.2. The Extrusion Process

The diameter of the extruded blank was machined to 59 mm prior to extrusion of the metal matrix composite. Under the temperature of 300 °C, the billet was extruded at a speed of 2 mm/s with an extrusion ratio of 30, and a half-angle of the extrusion die of 30°, as shown as in Figure 3a. The geometric model of extrusion die was shown in Figure 3b. The extruded material was immediately quenched to prevent the grain from growing.

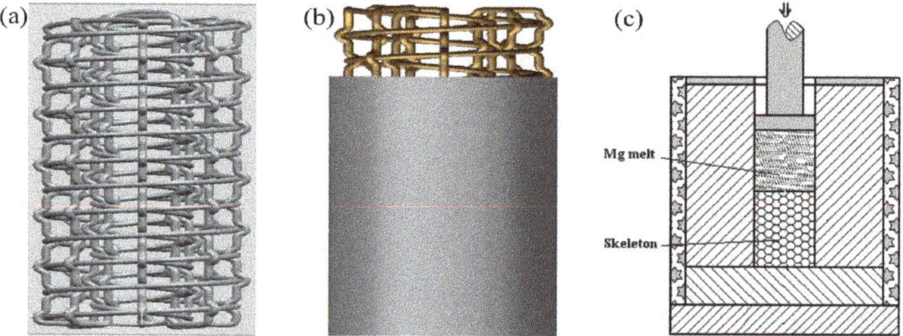

Figure 2. Three-dimensional model image and technical diagram: (**a**) reinforcement; (**b**) MIMC composites; (**c**) the pressure infiltration technology.

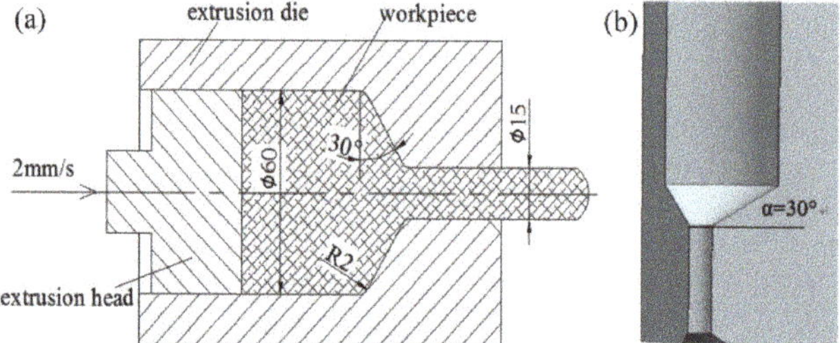

Figure 3. The schematic diagram of extrusion process: (**a**) pressure infiltration processing of the extrusion process; (**b**) 1/4 extrusion die of matrix composites interpenetrated by metal reinforcement (MIMC) composites.

2.3. Microstructure Characterization

The samples were mechanically polished using a polycrystalline diamond suspension ethylene glycol organic solution for observation of the optical microstructure (OM, Keyence VW-5000, Osaka, Japan). Then, the samples were subjected to etching in a solution consisting of acetic acid (5 mL), picric acid (5 g), ethanol (100 mL), and distilled water (10 mL), for 7 s, to reveal the crystal structure. The average grain size was analyzed by the Image Pro Plus (IPP, Media Cybernetics, Rockville, MD, USA). In addition, the EBSD samples were prepared by mechanical polishing using 2000# sandpaper and, then, by electric polishing in an electrolyte consisting of 400 mL butyl glycol ($C_4H_{10}O_2$, 99.5%), 80 mL ethanol (C_2H_6O, 95%), and 40mL perchloric acid ($HClO_4$, 70%), at a voltage of 11 V for 120 s. Microstructure and morphology of MIMC composites were investigated by scanning electron microscopy (SEM, Hitachi S-4800, Tokyo, Japan), and the elemental composition of the samples was determined by the energy dispersive spectroscopy (EDS, Oxford Instruments Inca Energy 350, Oxford, UK), as well as grain distribution and diameter measurements, which were determined using electron backscatter diffraction (EBSD, Oxford Instruments NordlysMax2, Oxford, UK).

2.4. Mechanical Properties Test

The extruded samples used for tensile testing were processed on a CNC lathe (CJK6130L, Shanghai, China), in accordance to the procedure outlined in American Society for Testing and Materials (ASTM E8M-96). There were no obvious processing marks on the surface of the test piece, nor were there any signs of breaking nucleation. The mechanical properties of the samples were tested using an Instron 5569 testing machine (Instron, Shanghai, China) with a crosshead moving speed of 0.1 mm/min, and the strain was monitored by a strain gage of 20 mm in length. At the same time, the test comparisons of the AZ31 magnesium alloy were carried out, and each tensile value was the average of at least three measurements.

3. Results

3.1. Microstructure

Figure 4 shows the surface microstructure and cross-section characteristics of MIMC composites. From Figure 4a, the matrix and reinforcement are interpenetrating, anisotropic, and interwoven. In addition, in the longitudinal direction, a portion of the structure of the interpenetrating network of reinforcements is exposed from the matrix (white circles), and another portion of the reinforcement is wrapped and buried by magnesium matrix. Only a few are exposed from the matrix in the transverse direction (black arrow). In addition, the matrix and reinforcement form a good interface combination, as shown as in Figure 4b. Furthermore, the distribution of the chemical elements of the samples are examined by EDS, and subjected to line scanning, as shown as in Figure 5. The reinforcements of the circular area are stainless steel (cyan line in Figure 5a), titanium alloy (blue line in Figure 5b), and aluminum alloy (cyan line in Figure 5c), respectively. The magnesium matrix is represented as a red line in the MISC and MIAC composites, however, it is described as a green line in MITC composite.

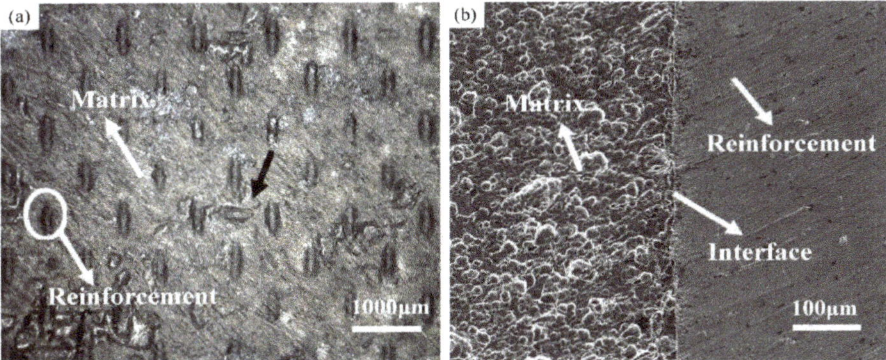

Figure 4. SEM micrograph of MIMC composites: (**a**) the surface microstructure; (**b**) cross-section of reinforcement with matrix.

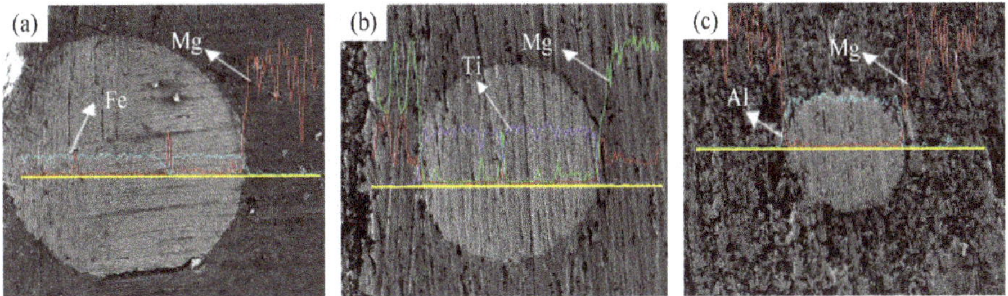

Figure 5. The energy dispersive spectroscopy (EDS) analysis of MIMC composites: (**a**) magnesium matrix composites interpenetrated by stainless steel reinforcement (MISC); (**b**) magnesium matrix composites interpenetrated by titanium alloy reinforcement (MITC); (**c**) magnesium matrix composites interpenetrated by aluminum alloy reinforcement (MIAC).

Figure 6 shows the OM microstructure morphology and grain size distribution of magnesium matrix in three kinds of composite as-cast states. In MIMC composites, their grain distribution and grain size have the same characteristics, that is, the grain distribution does not obey a normal distribution and is randomly distributed. In addition, the grain size is mainly concentrated in the range 200–300 μm. Therefore, the results show that the grain and its area fraction are not affected by any of the reinforcement materials. However, compared with the as-cast state (Figure 6a,c,e), the micromorphology of the as-extruded samples has greater changes, which are shown in Figure 7a,c,e. The relatively small grain size of the as-extruded samples is owing to the accumulation of shear strain and dynamic recrystallization during extrusion. However, the grain refinement degree is different in three kinds of composites. In MISC composites, it can be seen, from Figure 7a, that the grain size follows a normal distribution. Moreover, most of the grain size is 10–20 μm, so the average grain size is 14 μm. In MITC composites, although the grain size does not conform to the normal distribution (Figure 7d), most of them are concentrated between 15–30 μm. Therefore, the average grain size is 20 μm. In MIAC composites, the grain size adheres to the Weibull distribution (Figure 7f), and most of them are concentrated between 20–30 μm. The average grain size is 24 μm. In short, the different reinforcement materials result in different microstructure characterization of magnesium matrix during deformation. The reason is that, due to the different strengths of reinforcement, the magnesium matrix bears a different stress from the reinforcement during extrusion. As we know, magnesium matrix alloys usually form a strong texture, resulting in the decreases of ductility at room temperature because of their intrinsic characterization of the hexagonal close-packed structure [29]. Since the room temperature-forming properties of the magnesium alloys are mainly affected by the initial crystal structure of the basal plane, the improvement of the crystal grains will significantly improve the formability of the MIMC composites [30]. In MISC composites, the three-dimensional network of stainless-steel has the highest strength, which leads to greater stress on magnesium grains, thus causing a till of the basal plane during coordinated deformation, and a minimum average grain size. Therefore, the dynamic recrystallization occurs in the vicinity of the reinforcement of magnesium matrix composite. Furthermore, the difference between size distribution and grain refinement can be explained by the changes in the grain orientation of magnesium matrix composite by EBSD test. Figure 8a shows that the grain orientation of the as-cast magnesium matrix composite (MIMC) is almost randomly distributed. Figure 8b–d reveal the grain orientation changes of the as-extruded magnesium matrix composite. From Figure 8b (MISC), the results show that during the extrusion, the grain orientation forms a strong peak at the center of the basal pole figure of {0001}. However, in the MIAC composites, {0001} basal pole figure exhibits a divergence state, due to a relatively small stress for the aluminum alloy imposed to magnesium matrix, as shown as in Figure 8d. From Figure 8c (MITC), it is shown that the grain orientation is between MISC and MIAC.

3.2. Mechanical Properties

The MIMC composites exhibit different mechanical properties because of the difference of reinforcement materials. Figure 9 shows the mechanical properties change with different reinforcement materials. Compared with AZ31 magnesium alloy, the MIMC composites have a significant improvement in the yield strength (YS), ultimate tensile strength (UTS), and elongation (E). The improvement in mechanical performance is attributed to the following as (1) superior mechanical properties of the reinforcement itself; (2) the reinforcements have a good interface with the matrix; (3) effective load transfer can be carried out from matrix to reinforcement; and (4) fine and evenly distributed grain are produced during coordination deformation of the reinforcement and matrix.

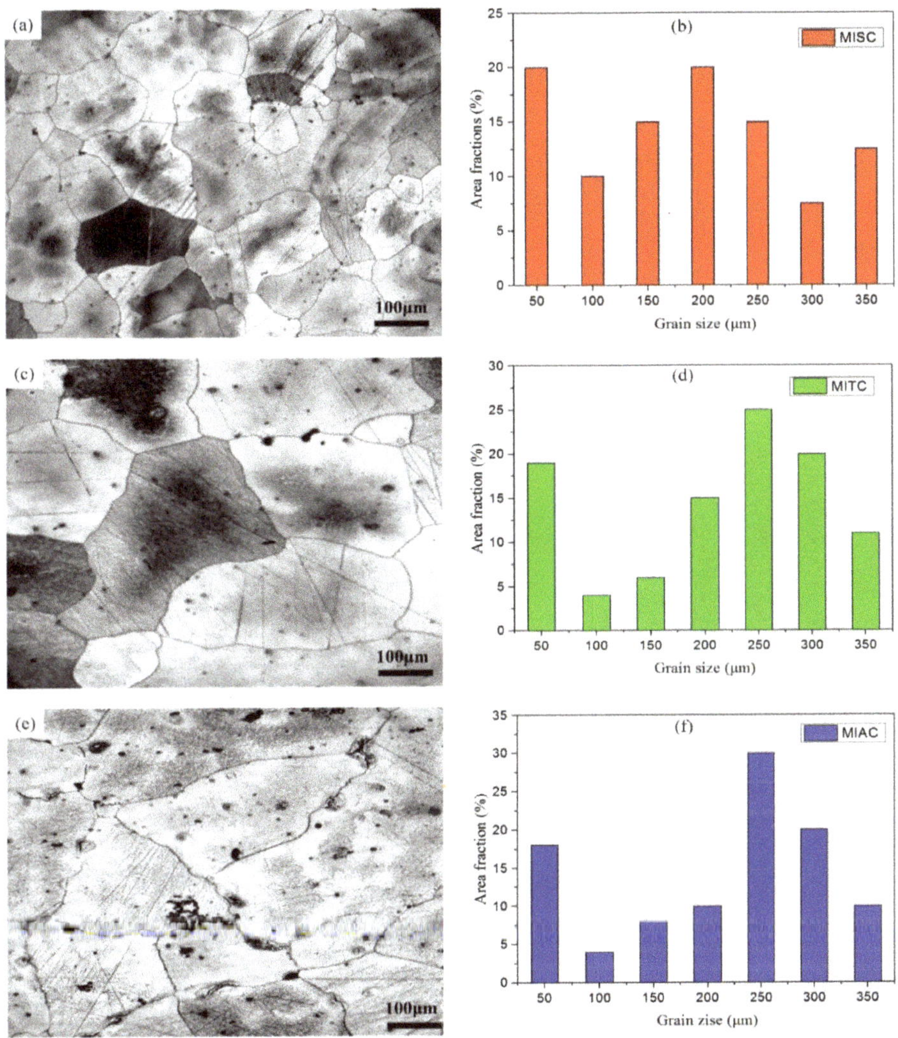

Figure 6. The optical microstructure and the grain size distributions of magnesium matrix composite in three kinds of reinforcement as-cast states: (**a**,**b**) MISC; (**c**,**d**) MITC; (**e**,**f**) MIAC.

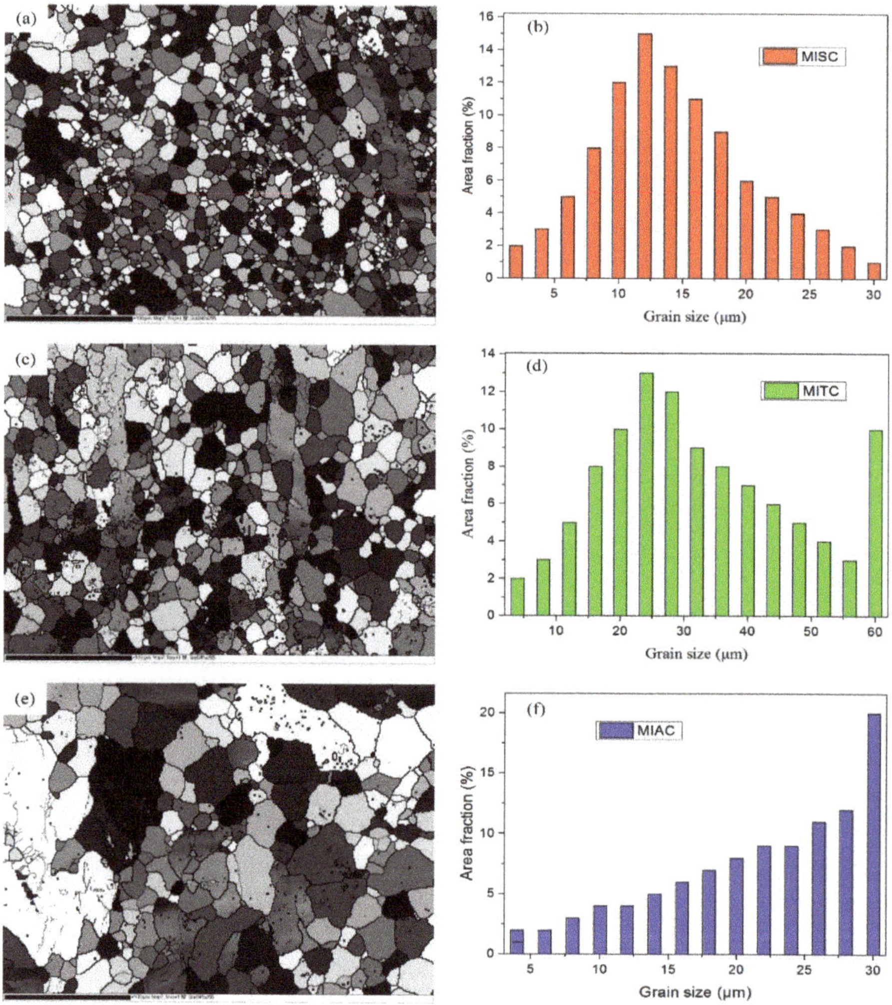

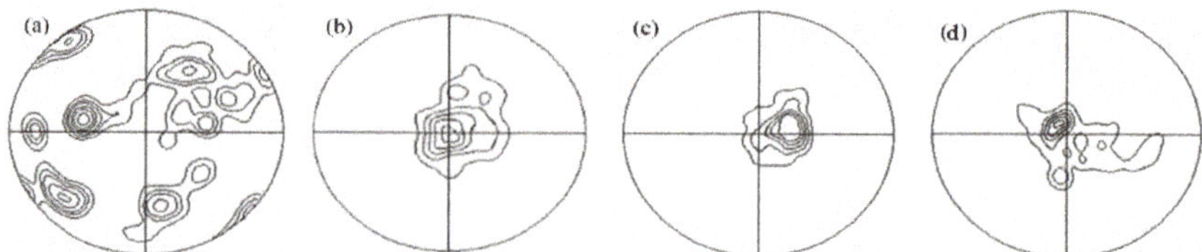

Figure 7. The SEM and the grain size distributions of magnesium matrix composite in three kinds of reinforcement as-extruded states: (**a**,**b**) MISC; (**c**,**d**) MITC; (**e**,**f**) MIAC.

Figure 8. The {0001} basal pole figures: (**a**) cast states of MIMC; (**b**) extruded state of MISC; (**c**) extruded state of MITC; (**d**) extruded state of MIAC.

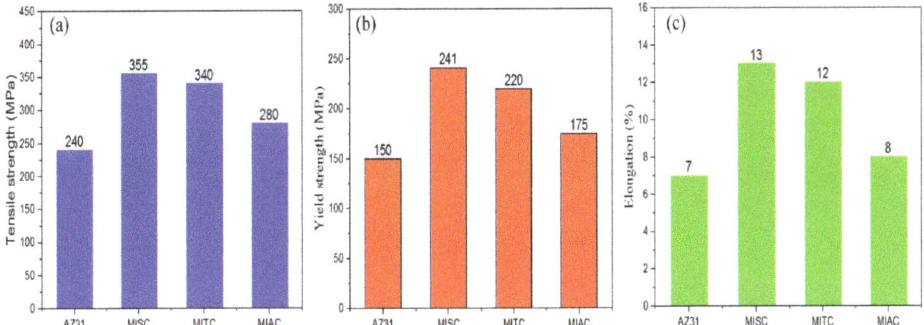

Figure 9. The mechanical properties of AZ31, MISC, MITC, and MIAC: (**a**) UTS; (**b**) YS; (**c**) E.

Among the three kinds of composites, the MISC composites have the maximum values of UTS (Figure 9a), YS (Figure 9b), and E (Figure 9c), which are 355 MPa, 241 MPa, and 13%. For each parameter, comparing with the AZ31, the increase is 47.9%, 60.7%, and 85.7%, respectively, while the MITC composites followed, the mechanical properties are 340 MPa, 220 MPa, and 12%, increasing by 41.7%, 46.7%, and 71.4%, respectively. The MIAC composites are the minimum, the mechanical properties of which are 280 MPa, 175 MPa, and 8%, increasing by 16.7%, 16.7%, and 14.3%, respectively. These changes are ascribed to two factors. One is the influence of reinforcement itself, and the mechanical properties of the three network reinforcements are shown in Table 2. Another is the influence of the grain refinement degree of matrix alloy.

Table 2. The mechanical properties of the three network reinforcements.

Materials	Tensile Strength (MPa)	Yield Strength (MPa)	Elongation (%)
Fe–18Cr–9Ni	893	824	36
Ti–6Al–4V	790	710	21
Al–5Mg–3Zn	542	520	11

4. Discussions

The reinforcement and matrix of MIMC composites are regarded as two interpenetrating pore structures or network structures, which are shown in Figure 10a. Owing to that one kind of metal phase is interpenetrated by another metal phase, the mechanical models of the network structure of MIMC composites does not follow the brittle fracture mode which is shown in Figure 10b, it follows the elastic plastic mode which is shown in Figure 10c, and this deformation mode is the main failure mode of MIMC composites [31,32]. Considering the linear elastic elements, the elastic modulus of interpenetrating composites is estimated as Equation (1), according to the mixing rule

$$E_C = E_M V_M + E_R V_R,$$ (1)

where E_C, E_M, and E_R are, respectively, the elastic modulus of composites (C), matrix (M), and reinforcement (R), and V_M and V_R are, respectively, the volume fraction of matrix and reinforcement.

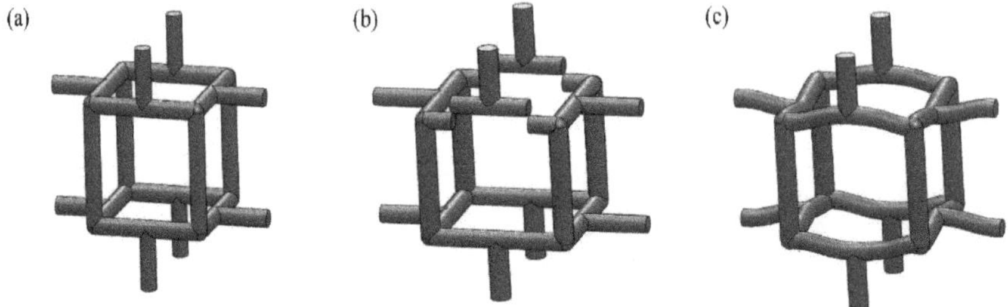

Figure 10. The mechanical model of network structure: (**a**) the unit model; (**b**) the brittle model; (**c**) the elastic plastic model.

Considering the elastic modulus influence, the constitutive mechanical mode of MIMC is proposed according to mixture law of composites, as shown in Equations (2) and (3).

$$\text{Upper limit model}: E_C X_C = \sum_{i=1}^{n} E_i X_i V_i, \tag{2}$$

$$\text{Lower limit model}: \frac{E_C}{X_C} = \sum_{i=1}^{n} \frac{E_i V_i}{X_i}, \tag{3}$$

where X_C represents the mechanical properties of composite materials, and X_i and V_i represent the mechanical properties and volume fractions of different composite phase, respectively. For two phases interpenetrating composites, the mechanical properties, such as tensile strength, yield strength, and elongation, are rewritten as Equations (4)–(7), according to the above equations,

$$E_C \sigma_C = E_M V_M \sigma_M + E_R V_R \sigma_R, \tag{4}$$

$$\frac{E_C}{\sigma_C} = E_M \frac{V_M}{\sigma_M} + E_R \frac{V_R}{\sigma_R}, \tag{5}$$

$$E_C \delta_C = E_M V_M \delta_M + E_R V_R \delta_R, \tag{6}$$

$$\frac{E_C}{\delta_C} = E_M \frac{V_M}{\delta_M} + E_R \frac{V_R}{\delta_R}, \tag{7}$$

where σ and V represent, respectively, the tensile strength or yield strength and volume fractions of the matrix (M) and reinforcement (R), and δ represents the elongation of the matrix (M) and reinforcement (R).

Combining Equation (1) to Equations (4)–(7), the following Equations (8)–(11) are obtained:

$$\text{Upper limit model}: \sigma_C = \frac{E_M}{E_M V_M + E_R V_R} V_M \sigma_M + \frac{E_R}{E_M V_M + E_R V_R} V_R \sigma_R, \tag{8}$$

$$\text{Lower limit model}: \frac{1}{\sigma_C} = \frac{E_M}{E_M V_M + E_R V_R} \frac{V_M}{\sigma_M} + \frac{E_R}{E_M V_M + E_R V_R} \frac{V_R}{\sigma_R}, \tag{9}$$

$$\text{Upper limit model}: \delta_C = \frac{E_M}{E_M V_M + E_R V_R} V_M \delta_M + \frac{E_R}{E_M V_M + E_R V_R} V_R \delta_R, \tag{10}$$

$$\text{Lower limit model}: \frac{1}{\delta_C} = \frac{E_M}{E_M V_M + E_R V_R} \frac{V_M}{\delta_M} + \frac{E_R}{E_M V_M + E_R V_R} \frac{V_R}{\delta_R}. \tag{11}$$

According to the above mathematics model, the relationship between volume fraction and tensile strength of magnesium matrix composites interpenetrated by different reinforcement is shown in Figure 11. The results show that as the volume fraction of reinforcement increases, the continuous improving trend of tensile strength is more obvious. Similarly, the yield strength and elongation also have the same regularity. Compared with the tested results in Figure 9, the results of the mathematics model are the same. Therefore, it can be considered that the mathematics model of the composite material, derived from this paper, is correct and provides a good theoretical guidance for the test.

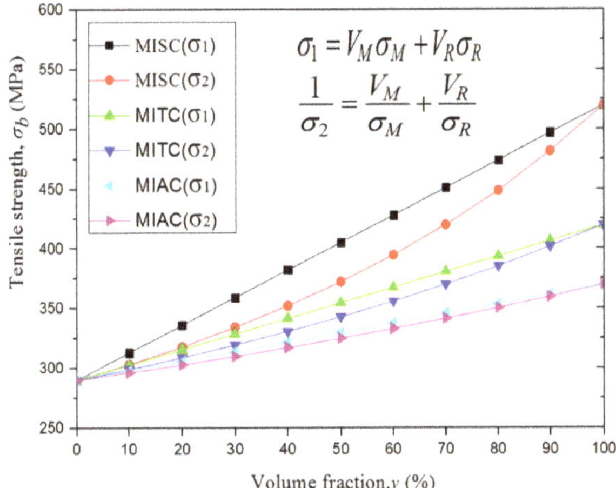

$$\sigma_1 = V_M\sigma_M + V_R\sigma_R$$
$$\frac{1}{\sigma_2} = \frac{V_M}{\sigma_M} + \frac{V_R}{\sigma_R}$$

Figure 11. The relationship between tensile strength and volume fraction of magnesium matrix composites interpenetrated by different reinforcement.

5. Conclusions

1. The magnesium matrix composites reinforced by stainless steel (Fe–18Cr–9Ni), titanium alloy (Ti–6Al–4V), and aluminum alloy (Al–5Mg–3Zn), were prepared by the pressure infiltration technology. The reinforcements of the composites are interwoven with the matrix and have an integrated interface.

2. The grain size and distribution in the as-cast magnesium matrix composites have the same characteristics, and their sizes are mainly concentrated at 200–300 μm, and the size distribution is random and does not obey normal distribution. However, during the extrusion process, the grain size of magnesium matrix composite is more refined than that of the as-cast state, because of the dynamic recrystallization and the superposition of shear strain, and its size is mostly concentrated at 10–30 μm. In addition, the grain size distributions of MISC and MIAC follow the normal distribution and the Wien distribution, respectively.

3. Different reinforcement materials result in different microstructure characterization of magnesium matrix during extrusion. The average grain sizes of magnesium matrix composites with the reinforcements of MISC, MITC, and MIAC are 14, 20, and 24 μm, respectively.

4. Compared with AZ31, the mechanical properties of metal composites interpenetrated by metal reinforcement have a significant improvement. Different MIMCs with different reinforcements exhibit different mechanical properties. Among the three composite materials, the tensile strength, yield strength, and elongation of MISC are 355 MPa, 241 MPa, and 13%, with an increase of 47.9%, 60.7%, and 85.7%, respectively; while for MITC, they are 340 MPa, 220 MPa, and 12%, with an increase of 41.7%, 46.7%, and 71.4%, respectively; and for MIAC, are 280 MPa, 175 MPa, and 8%, an increase of 16.7%, 16.7%, and 14.3%, respectively.

Author Contributions: S.W. (Shouren Wang) wrote the main part of the manuscript and developed the planning of the experiment. S.W. (Shuxu Wu) and Y.W. carried out the preparation of different composite materials and tested the mechanical properties. D.W. and G.W. performed characterization of the microstructure of the samples. S.W. (Shouren Wang) summed up the article and conducted a final review. S.W. (Shuxu Wu) made the final typesetting of the article.

Funding: This research was funded by the National Natural Science Foundation of China (No. 51872122), Shandong Key Research and Development Plan, China (No.: 2017GGX30140, 2016JMRH0218) Shandong Provincial Natural Science Foundation, China (No.: ZR2017BEE055), Distinguished Middle-Aged and Young Scientist Encourage and Reward Foundation of Shandong Province (No.: ZR2016EMB01) and Taishan Scholar Engineering Special Funding (2016–2020).

Conflicts of Interest: The authors declare no conflict of interest.

Nomenclature

MMCs	metal matrix composites
MIMC	metal matrix composites interpenetrated by metal reinforcement
MISC	magnesium matrix composites interpenetrated by stainless steel reinforcement
MITC	magnesium matrix composites interpenetrated by titanium alloy reinforcement
MIAC	magnesium matrix composites interpenetrated by aluminum alloy reinforcement
WR	warm rolling
TS	torsion straining
TRC	twin-roll casting
RE	reciprocal extrusion
OM	optical microstructure
SEM	scanning electron microscopy
EDS	energy dispersive spectroscopy
EBSD	electron backscatter diffraction
CNC	computer numerical control
YS	yield strength
UTS	ultimate tensile strength
E	elongation
E_C, E_M and E_R	the elastic modulus of composites, matrix, and reinforcement
V_M and V_R	the volume fraction of matrix and reinforcement
X_C	the mechanical properties of composite materials
X_i and V_i	the mechanical properties and volume fractions of different composite phase
σ and V	the tensile strength or yield strength and volume fractions of the matrix and reinforcement
δ	the elongation of the matrix and reinforcement

References

1. Mortensen, A.; Llorca, J. Metal matrix composites. *Annu. Rev. Mater. Res.* **2010**, *40*, 243–270. [CrossRef]
2. Nicholls, C.J.; Boswell, B.; Davies, I.J.; Islam, M.N. Review of machining metal matrix composites. *Int. J. Adv. Manuf. Technol.* **2017**, *90*, 2429–2441. [CrossRef]
3. Kim, C.S.; Cho, K.; Manjili, M.H.; Nezafati, M. Mechanical performance of particulate-reinforced Al metal-matrix composites (MMCs) and Al metal-matrix nano-composites (MMNCs). *J. Mater. Sci.* **2017**, *52*, 13319–13349. [CrossRef]
4. Tevatia, A.; Srivastava, S.K. Modified shear lag theory based fatigue crack growth life prediction model for short-fiber reinforced metal matrix composites. *Int. J. Fatigue* **2015**, *70*, 123–129. [CrossRef]
5. Roger, J.; Gardiola, B.; Andrieux, J.; Viala, J.C.; Dezellus, O. Synthesis of Ti matrix composites reinforced with TiC particles: Thermodynamic equilibrium and change in microstructure. *J. Mater. Sci.* **2017**, *52*, 4129–4141. [CrossRef]
6. Akatsu, T.; Takashima, H.; Shinoda, Y.; Wakai, F.; Wakayama, S. Thermal-shock fracture and damage resistance improved by whisker reinforcement in alumina matrix composite. *Int. J. Appl. Ceram. Technol.* **2016**, *13*, 653–661. [CrossRef]
7. Leclerc, W.; Ferguen, N.; Pelegris, C.; Haddad, H.; Bellenger, E.; Guessasma, M. A numerical investigation of effective thermoelastic properties of interconnected alumina/Al composites using FFT and FE approaches. *Mech. Mater.* **2016**, *92*, 42–57. [CrossRef]
8. Li, S.; Xiong, D.G.; Liu, M.; Bai, S.X.; Zhao, X. Thermophysical properties of SiC/Al composites with three dimensional interpenetrating network structure. *Ceram. Int.* **2014**, *40*, 7539–7544. [CrossRef]

9. Al-Ketan, O.; Assad, M.A.; Al-Rub, R.K.A. Mechanical properties of periodic interpenetrating phase composites with novel architected microstructures. *Compos. Struct.* **2017**, *176*, 9–19. [CrossRef]
10. Cheng, F.F.; Kim, S.M.; Reddy, J.N.; Al-Rub, R.K.A. Modeling of elastoplastic behavior of stainless-steel/bronze interpenetrating phase composites with damage evolution. *Int. J. Plasticity* **2014**, *61*, 94–111. [CrossRef]
11. Yao, B.B.; Zhou, Z.Y.; Duan, L.Y.; Chen, Z.T. Anisotropic charpy impact behavior of novel interpenetrating phase composites. *Vacuum* **2018**, *155*, 83–90. [CrossRef]
12. Hidalgo-Manrique, P.; Yan, S.; Lin, F.; Hong, Q.; Kinloch, I.A.; Chen, X.; Young, R.J.; Zhang, X.Y.; Dai, S.L. Microstructure and mechanical behaviour of aluminium matrix composites reinforced with graphene oxide and carbon nanotubes. *J. Mater. Sci.* **2017**, *52*, 13466–13477. [CrossRef]
13. Boyer, C.; Figueiredo, L.; Pace, R.; Lesoeur, J.; Rouillon, T.; Visage, C.L.; Tassin, J.-F.; Weiss, P.; Guicheux, J.; Rethore, G. Laponite nanoparticle-associated silated hydroxypropylmethyl cellulose as an injectable reinforced interpenetrating network hydrogel for cartilage tissue engineering. *Acta Biomater.* **2018**, *65*, 112–122. [CrossRef] [PubMed]
14. Swain, M.V.; Coldea, A.; Bilkhair, A.; Guess, P.C. Interpenetrating network ceramic-resin composite dental restorative materials. *Dent. Mater.* **2016**, *32*, 34–42. [CrossRef] [PubMed]
15. Kádár, C.; Máthis, K.; Knapek, M.; Chmelík, F. The effect of matrix composition on the deformation and failure mechanisms in metal matrix syntactic foams during compression. *Materials* **2017**, *10*, 196. [CrossRef] [PubMed]
16. Lichtenberg, K.; Weidenmann, K.A. Effect of reinforcement size and orientation on the thermal expansion behavior of metallic glass reinforced metal matrix composites produced by gas pressure infiltration. *Thermochim. Acta* **2017**, *654*, 85–92. [CrossRef]
17. Alhajeri, S.N.; Al-Fadhalah, K.J.; Almazrouee, A.I.; Langdon, T.G. Microstructure and microhardness of an Al-6061 metal matrix composite processed by high-pressure torsion. *Mater. Charact.* **2016**, *118*, 270–278. [CrossRef]
18. Beranoagirre, A.; Urbikain, G.; Calleja, A.; Lacalle, L.L.D. Hole making by electrical discharge machining (EDM) of γ-TiAl intermetallic alloys. *Metals* **2018**, *8*, 543. [CrossRef]
19. Thompson, A.; Senin, N.; Maskery, I.; Leach, R. Effects of magnification and sampling resolution in X-ray computed tomography for the measurement of additively manufactured metal surfaces. *Precis. Eng.* **2018**, *53*, 54–64. [CrossRef]
20. Tokutomi, J.; Uemura, T.; Sugiyama, S.; Shiomi, J.; Yanagimoto, J. Hot extrusion to manufacture the metal matrix composite of carbon nanotube and aluminum with excellent electrical conductivities and mechanical properties. *CIRP Ann.* **2015**, *64*, 257–260. [CrossRef]
21. Xin, L.; Yang, W.S.; Zhao, Q.Q.; Dong, R.H.; Liang, X.; Xiu, Z.Y.; Hussain, M.; Wu, G.H. Effect of extrusion treatment on the microstructure and mechanical behavior of SiC nanowires reinforced Al matrix composites. *Mater. Sci. Eng. A* **2017**, *682*, 38–44. [CrossRef]
22. Srikanth, N.; Kurniawan, L.A.; Gupta, M. Effect of interconnected reinforcement and its content on the damping capacity of aluminium matrix studied by a new circle-fit approach. *Compos. Sci. Technol.* **2003**, *63*, 839–849. [CrossRef]
23. Thakur, S.K.; Gupta, M. Use of interconnected reinforcement in magnesium for stiffness critical applications. *Mater. Sci. Technol.* **2008**, *24*, 213–220. [CrossRef]
24. Mondet, M.; Barraud, E.; Lemonnier, S.; Guyon, J.; Allain, N.; Grosdidier, T. Microstructure and mechanical properties of AZ91 magnesium alloy developed by spark plasma sintering. *Acta Mater.* **2016**, *119*, 55–67. [CrossRef]
25. Ghasali, E.; Alizadeh, M.; Shirvanimoghaddam, K.; Mirzajany, R.; Niazmand, M.; Faeghi-Nia, A.; Ebadzadeh, T. Porous and non-porous alumina reinforced magnesium matrix composite through microwave and spark plasma sintering processes. *Mater. Chem. Phys.* **2018**, *212*, 252–259. [CrossRef]
26. Wang, S.R.; Sun, B.; Geng, H.R.; Wang, Y.Z. The abrasive wear properties of Al-Mg-Si$_3$N$_4$ metal matrix composites. *J. Mater. Eng. Perform.* **2006**, *15*, 549–552.
27. Wang, S.R.; Geng, H.R.; Song, B.; Wang, Y.Z. Machinability of metal matrix composites reinforced by 3-D network structure. *Appl. Compos. Mater.* **2006**, *13*, 385–395. [CrossRef]
28. Wang, S.R.; Geng, H.R.; Wang, Y.Z. Fabrication and machinability of Si$_3$N$_4$-Mg-Al-Zn (AZ91) composites. *Mater. Sci. Technol.* **2006**, *22*, 223–226. [CrossRef]

29. Arkhurst, B.M.; Lee, M.Y.; Kim, J.H. Effect of resin matrix on the strength of an AZ31 Mg alloy-CFRP joint made by the hot metal pressing technique. *Compos. Struct.* **2018**, *201*, 303–314. [CrossRef]

30. Jiang, M.G.; Xu, C.; Yan, H.; Fan, G.H.; Nakata, T.; Lao, C.S.; Chen, R.S.; Kamado, S.; Han, E.H.; Lu, B.H. Unveiling the formation of basal texture variations based on twinning and dynamic recrystallization in AZ31 magnesium alloy during extrusion. *Acta Mater.* **2018**, *157*, 53–71. [CrossRef]

31. Mishra, A.; Mahesh, S. A deformation-theory based model of a damaged metal matrix composite. *Int. J. Solids Struct.* **2017**, *121*, 228–239. [CrossRef]

32. Trzepieciński, T.; Ryzińska, G.; Gromada, M.; Biglar, M. 3D microstructure-based modelling of the deformation behaviour of ceramic matrix composites. *J. Eur. Ceram. Soc.* **2018**, *38*, 2911–2919. [CrossRef]

Article

Characterization of Copper–Graphite Composites Fabricated via Electrochemical Deposition and Spark Plasma Sintering

Myunghwan Byun [1,*], Dongbae Kim [2], Kildong Sung [2], Jaehan Jung [3], Yo-Seung Song [4], Sangha Park [2,*] and Injoon Son [5,*]

1   Department of Advanced Materials Engineering, Keimyung University, Daegu 42601, Korea
2   Advanced Materials Research Group, Daegu Mechatronics and Materials Institute (DMI), Daegu 42714, Korea
3   Department of Materials Science and Engineering, Hongik University, Sejong 30016, Korea
4   Department of Materials Engineering, Korea Aerospace University, Goyang 10540, Korea
5   Department of Materials Science and Metallurgical Engineering, Kyungpook National University (KNU), Daegu 41566, Korea
*   Correspondence: myunghbyun@kmu.ac.kr (M.B.); shpark@dmi.re.kr (S.P.); ijson@knu.ac.kr (I.S.); Tel.: +82-53-580-5228 (M.B.); +82-53-608-2131 (S.P.); +82-53-950-5563 (I.S.); Fax: +82-53-580-5165 (M.B.); +82-53-608-2119 (S.P.); +82-53-950-6659 (I.S.)

Received: 27 June 2019; Accepted: 15 July 2019; Published: 17 July 2019

Abstract: In the present study, we have demonstrated a facile and robust way for the fabrication of Cu-graphite composites (CGCs) with spatially-aligned graphite layers. The graphite layers bonded to the copper matrix and the resulting composite structure were entirely characterized. The preferential orientation and angular displacement of the nano-sized graphite fiber reinforcements in the copper matrix were clarified by polarized Raman scattering. Close investigation on the change of the Raman G-peak frequency with the laser excitation power provided us with a manifestation of the structural and electronic properties of the Cu-graphite composites (CGCs) with spatially-distributed graphite phases. High resolution transmission electron microscopy (TEM) observation and Raman analysis revealed that reduced graphite oxide (rGO) phase existed at the CGC interface. This work is highly expected to provide a fundamental way of understanding how a rGO phase can be formed at the Cu-graphite interface, thus finally envisioning usefulness of the CGCs for thermal management materials in electronic applications.

Keywords: copper–graphite composites; anisotropic layered structure; spark plasma sintering; Raman spectroscopy

---

1. Introduction

As the trend in power electronics systems moves toward smart and wearable electronics, effective use of thermal management materials is of critical importance due to the strong demand for the enhancement of power densities and miniaturization and weight reduction [1–6]. Of the various thermal management elements, copper (Cu) is among the most widely used materials for various kinds of passive heat exchangers including sinks, spreads, and pipes because of its excellent thermal conduction (~400 W/m·K) and weldability with solders, thereby making it compatible with electronic applications [7]. However, since Cu has a relatively high density of 8.96 g/cm^3 compared to aluminum (2.70 g/cm^3), its use can be restricted in applications such as the heat dissipation of electronics, automobile and aerospace electronics, of which space and weight is a strict requirement [3,8,9]. Therefore, fabricating and developing a new heat management materials system with lower density and good weldability still remains a big challenge in future electronic cooling.

Recently, as a promising alternative, carbon allotropes such as diamond [10], natural graphite [11], synthetic graphite [12,13], carbon nanotubes/fiber/flakes [14], and graphene [15–18] have attracted great attention. However, when a polymeric thermal adhesive is applied for passively exchanging the heat at the interfacial region between the heat source and the carbon-based heat dissipater, undesired thermal stress and warpage in electronic components possibly takes place because of discrepancy in thermal expansion coefficients [3,19,20]. Another presumable issue in the solely used carbon-based materials is weak bonding with solders, thus leading to ascending thermal resistance [3,19,20]. To amend the aforementioned issues, effort to combine Cu with carbon-based materials have been made to enhance thermal management capability and concurrently reduce the density of Cu [21,22]. Actually, recent studies have successfully shown a variety of routes to fabrication of Cu-matrix composites by incorporating the carbon allotropes into a Cu matrix such as hot pressing [22], vacuum pressure infiltration [21], electro- [3,23,24] and electroless plating [25], chemical vapor deposition [26], spark plasma sintering [27], etc. Unfortunately, most studies have focused on improvement of thermal conductivity, reduction of Cu density and the distribution of carbon-based reinforcements, rather than characterization of the interfacial region between the Cu matrix and the carbon allotropes as reinforcements.

In the present study, we have demonstrated a facile and robust way for the fabrication of Cu-graphite composites (CGCs) with spatially-positioned graphite phases. The graphite layers bonded to the copper matrix and the resulting composite structure were entirely characterized. The preferential orientation and angular displacement of the nano-sized graphite fiber reinforcements in the copper matrix were clarified by polarized Raman scattering. Close investigation of the change of the Raman G-peak frequency with the laser excitation power provided us with a manifestation of the structural and electronic properties of the Cu-graphite composites (CGCs) with spatially-distributed graphite phases. Here, we compared the thermal characteristics of *G*-peak shifts and strengths due to thermal reduction at the copper, graphite, and graphene interfaces in the composites.

2. Experimental Procedure

2.1. Preparation of Copper–Graphite Composite Materials

Graphite (fiber type, density of ~2.2 g/cm^3, Qingdao Krofmuehl Graphite Co., Ltd., Pingdu, China) with a size discrepancy ranging from 100 to 120 μm was selected as the reinforcement material. Prior to electroless plating of Cu on the whole surface of the graphite fiber, the graphite fibers were thermally handled at an elevated temperature of 380 °C in air for 60 min to activate the surface and then the samples were ultra-sonicated in acetic acid (CH$_3$CO$_2$H). The Cu coating on the graphite fibers was conducted in an electroless plating bath containing an aqueous solution of 70 wt.% cupric sulfate pentahydrate (CuSO$_4$·5H$_2$O) and 10 wt.% formaldehyde (HCHO) at 45 °C with pH values of 8–11 (tuned with a varying content of NaOH). The coating thickness of Cu on graphite fibers was determined with the graphite fraction added to the electroless plating bath. After vigorously rinsing the samples with distilled water, they were dried in a vacuum oven at 60 °C. Finally, the graphite fibers coated with Cu of 2–3 μm were obtained.

2.2. Spark Plasma Sintering

Consolidation of copper-coated graphite samples was performed as follows: First, the samples were loaded into a rectangular graphite die (inner diameter of 40 mm × 40 mm) and then thermally treated by using spark plasma sintering (SPS-3.20MK-V, Dr. Sinter, Sumitomo Coal Mining Co., Ltd., Saitama, Japan) under a controlled condition of pressure ~50 MPa and temperature ~920 °C, thereby forming the copper–graphite composites (CGCs).

2.3. Characterization of Copper–Graphite Composites

A Raman spectroscopy (InVia Reflex, Renishaw) was applied for characterizing the chemical structure of the carbon in the composites. A characteristic laser wavelength of 514 nm was used to irradiate the surface of the composite material. The interfacial area was carefully observed by using field-emission scanning electron microscopy (FE-SEM; JSM-7900F, JEOL) and transmission electron microscopy (TEM; JEM-2100F, JEOL). The chemical composition of the copper-graphite interface in the copper-graphite composite sintered body was intensively investigated by X-ray photoelectron spectroscopy (XPS; ESCALAB 250).

3. Results and Discussion

The graphite fibers were coated with Cu through an electroless plating process, thus forming the CGC powders with a volume ratio of 30 wt.% graphite and 70 wt.% Cu as shown in Figure 1a,b. Thickness of the deposited Cu was observed to be within the range of 2 to 3 μm. Subsequently, these composite powders were consolidated during the spark plasma sintering (SPS) process as displayed in Figure 1c,d. Cross-sectional images of the sintered CGCs uncovered that the graphite powders were layered in the Cu matrix, representing anisotropy in the mixed phase. The graphite layers (black colored region) were spatially distributed in the Cu matrix (white colored region), thus aligning in the plane direction perpendicular to the pressing direction during the sintering process. To clearly identify interfacial reconstruction in between the Cu matrix and the graphite oxide layers and, more importantly, to rationalize the structure formation driven during the SPS process, cross-sectional TEM observations were performed. Such these significant structural changes were made from pristine graphite to graphite oxide (GO), and to the reduced graphite oxide (rGO), which took place during the SPS process. Previous studies of rGO phases formed in the Cu matrix have focused on discussing mechanical properties dominated by the interface strength between Cu and graphene and the graphene dispersion in the composite [28–30]. It is worth noting that the copper–graphite interface with a well-defined thickness of the graphite oxide layers ranging from 8 to 10 nm were formed as obviously shown in Figure 2. Figure 3 shows the XPS results for the chemical composition and ionization etch depth at the copper–graphite composite interface. The Cu and Cu_2O phases, and the composition of carbon, oxygen, and copper compounds was confirmed to be 10 nm from the depth profile of 40 nm on the surface. It could be seen that the composition and quantitative changes in carbon, oxygen, and copper compounds were found to be 20 nm in the depth direction from the surface of the composite material. The amount of carbon and oxygen atoms decreased with distance from the surface, while the pure copper composition increased in weight. This strongly reflected that the rGO and GO interfacial layers formed the anisotropic composite structure [31–33].

Structural changes of the CGCs during the SPS process were further investigated using Raman spectroscopy. The Raman spectrum of the CGC powder showed a G band shift at 1581 cm^{-1} corresponding to the primary scattering of graphite, where the intensity was higher than that of the D band as displayed in Figure 4. The Raman spectrum of the interfacial graphite and Cu after the SPS process showed a weak intensity peak at 1594 cm^{-1}, as well as the G band at 1581 cm^{-1}. In addition, the intensity of the D band at 1363 cm^{-1} increased remarkably. These observations were due to the co-existence of the graphite oxide and graphite phases; hence, we reached the conclusion that the graphite surface was highly expected to be oxidized during the electroless plating and the SPS processes. Furthermore, the increased intensities in G and D bands (1594 and 1352 cm^{-1}, respectively) originated from the formation of GO [34,35]. Since GO and pristine graphite are the different allotropes of carbon, the graphite oxide surface can also be readily bonded to an oxygen-containing group, thereby resulting in ascending distance between the carbon atoms. Therefore, we concluded that the weak peak at 1594 cm^{-1} observed in the Raman spectrum of the CGC material interface turned up after the SPS process [8,36].

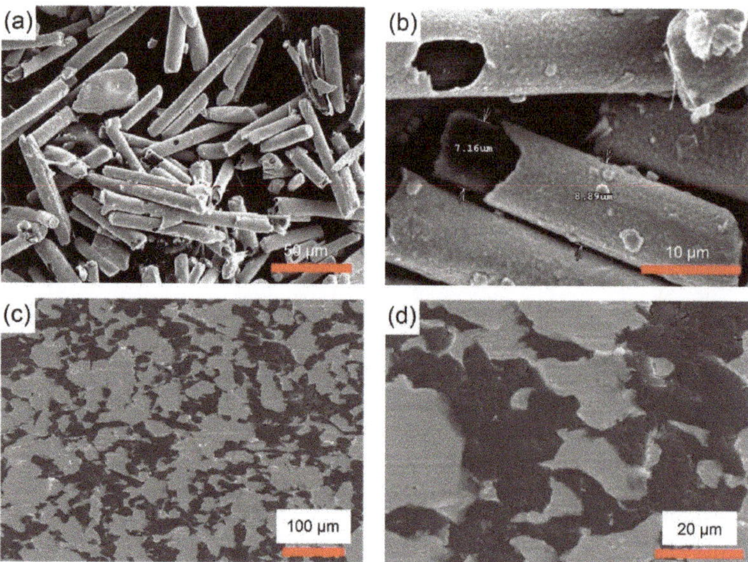

Figure 1. Field emission electron microscopy (FE-SEM) images of Cu-coated graphite fibers (**a,b**) and Cu-graphite composites after the spark plasma sintering process (**c,d**).

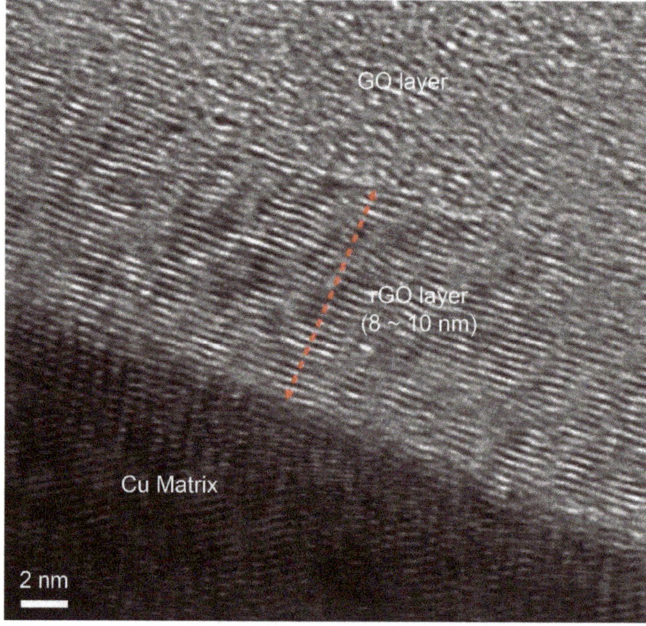

Figure 2. High-resolution transmission electron microscopy image of the interface between the copper and graphite composite.

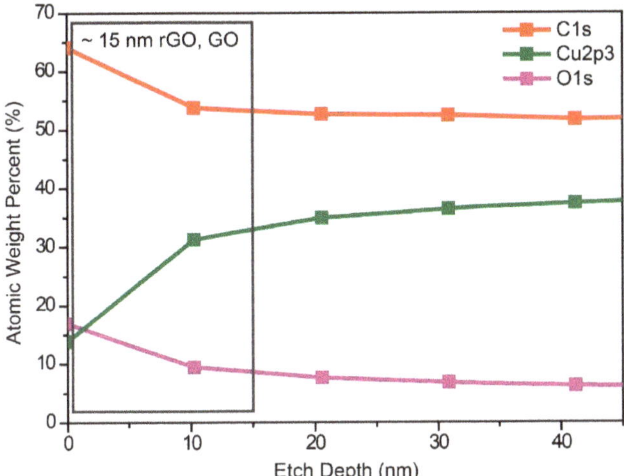

Figure 3. X-ray photoelectron spectroscopy of the interface between the copper and graphite composite.

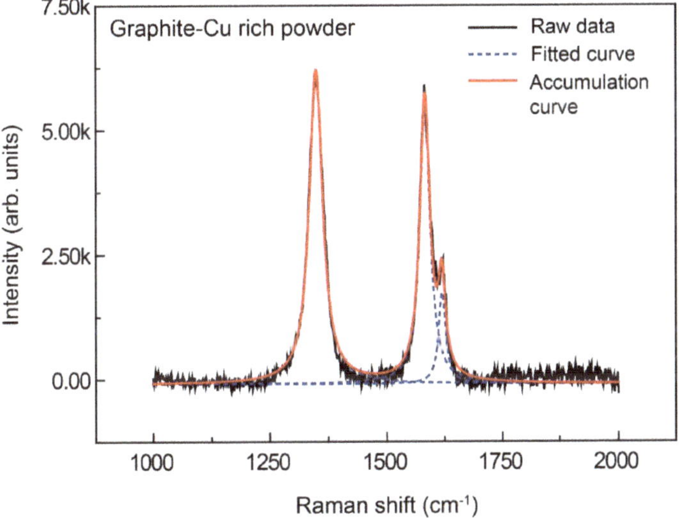

Figure 4. Comparison of Raman spectra for the Cu-coated graphite powders, copper–graphite composite interface, and Cu surface.

For further clarification, Raman mapping was performed over the interfacial positions of the copper–graphite interface equivalent to the graphite oxide, reduced graphite powder, and copper–graphite sintered body as clearly marked in Figure 5a. Five representative regions indicate X_1 (~1 μm), X_2 (~6 μm), X_3 (~15 μm), and X_4 (~28 μm), which corresponded to the distance from the composite interface, respectively. The D band indicates the presence of disordered carbons, whereas the G band indicates the graphitized carbon. Therefore, peak changes in D and G bands over five mapping regions reflect the existence probability of the oxygenic functional groups and disordered carbons on the CGC interface after the SPS process. Particularly, Raman spectra measured at X_3 (~15 μm) meant a lower presence probability of the oxygenic functional groups and disordered carbons compared with other three regions. This assumption could be reasonably rationalized by the fact that GO tends to be readily reduced to graphene-like sheets by getting rid of the oxygenic functional groups with the restoration of a π-conjugated structure.

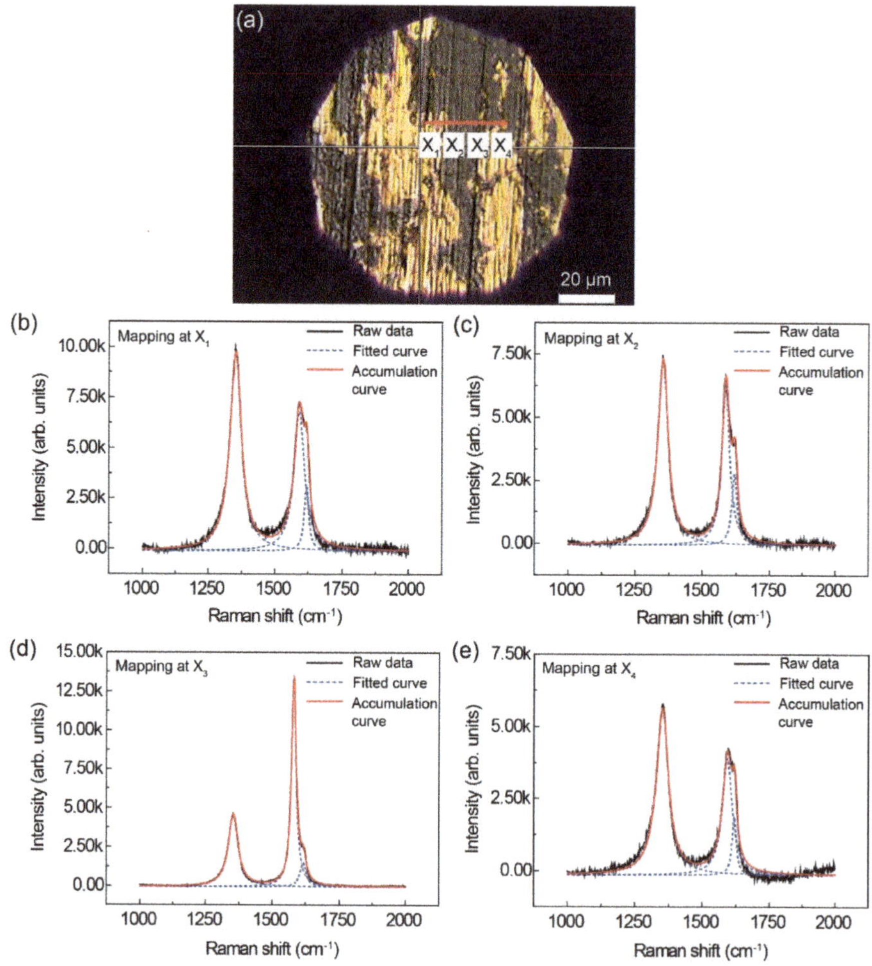

Figure 5. Raman mapping was performed over the interfacial positions of the copper–graphite interface equivalent to the graphite oxide, reduced graphite powder, and copper–graphite sintered body. (**a**) A cross-sectional view of the copper-graphite composite visualized by optical microscopy. Raman mapping results for five representative spots of the Cu-graphite composites interface (**b**) 1, (**c**) 6, (**d**) 15, and (**e**) 28 μm from the composite interface.

The Raman shift (cm^{-1}) and intensity ratio (I_D/I_{G1}) were introduced to firmly confirm the presence probability of rGO at the composite interface as plotted in Figure 6. As a result, we concluded that rGO was formed at the copper–graphite interface [37,38]. In the case of I_D/I_{G1}, analysis of rGO in the copper–graphite interface could not be easily handled because of inconsistence of the values for graphite, GO, and rGO with the mapping results. Characteristic peaks in the Raman spectra varied as the GO underwent chemical reduction. The peak around 1581 cm^{-1} was attributed to first-order scattering of phonons of *sp*2 carbon atoms, which is generally labeled as the *G* peak; similarly, a breathing mode of k-point photons of symmetry around 1348 cm^{-1} is referred to as the *D* peak. The I_D/I_{G1} ratio is designated as the degree of disorder, such as that due to defects, ripples, and edges. Figure 6 shows typical analytical data for graphite, GO, and the chemically rGO surface. The I_D/I_{G1} ratios were found to be 0.02, 0.7, and 0.87, respectively. The higher ratio of *D* to *G* bands strongly imply the higher presence probability of oxygenic functional groups and disordered carbons.

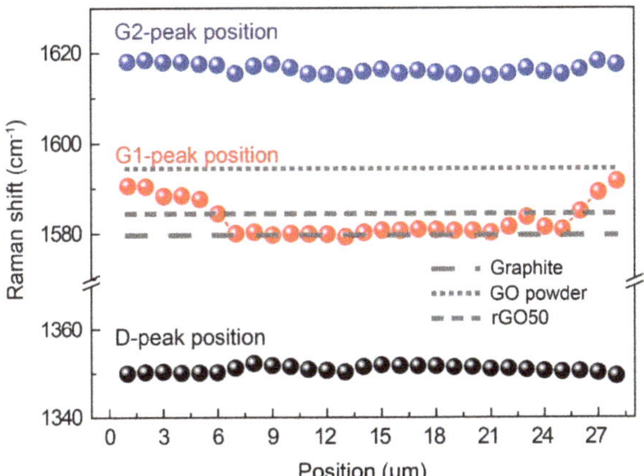

Figure 6. Comparison of the *G2-*, *G1-*, and *D*-peak positions and Raman shift (cm^{-1}) at the composite interface.

Finally, we investigated the thermal conductivity, conductivity, specific heat of the CGC along the through- and in-plane directions as displayed in Figure 7. All of the thermal properties of the CGC showed an obvious anisotropy, thus leading to 4 to 5 times higher thermal properties in the in-plane direction compared to the through-plane direction. These different tendencies along the through- and in-plane directions were predicted by applying the Hatta Taya method, which mainly assumes that the reinforcements are primarily oriented thin portions homogeneously dispersed in the matrix [39]. In an anisotropic structure, the phonon velocity can be varied along the through- and in-plane directions. However, the interfacial region with micrometer scale roughness could possibly take an effective crystallographic direction for energy propagation across the interface.

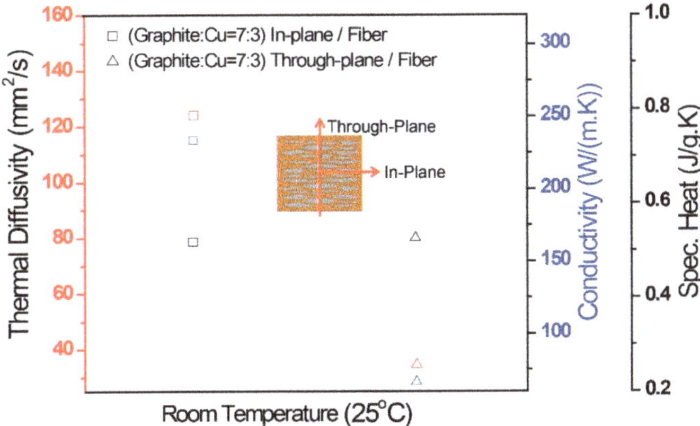

Figure 7. Thermal properties of the CGC along the through- and in-plane directions. Inset schematic represents both through- and in-plane directions of the CGC.

4. Conclusions

In summary, we demonstrated a facile and robust strategy to fabricate Cu-graphite composites (CGCs) with spatially-aligned, anisotropic layered structures through electroless deposition of graphite reinforcements with Cu and subsequent spark plasma sintering (SPS). On consolidation of the CGCs during the SPS process, rGO was formed at an interfacial region between the Cu matrix and the graphite layers. The formation of unprecedented rGO phases were intensively characterized by FE-SEM, TEM, XPS, and Raman spectroscopy, thus proving the presence probability of rGO phase experimentally. High resolution TEM observation and Raman analysis revealed that rGO phase existed at the CGC interface. This work is highly expected to provide a fundamental way of understanding how rGO phase can be formed at the Cu-graphite interface, thus finally envisioning usefulness of the CGCs for thermal management materials in electronic applications.

Author Contributions: M.B., and S.P. designed research; D.K., K.S., J.J., Y.-S.S., and S.P. performed research; M.B., and S.P. analyzed data; and S.P., I.S., and M.B. wrote the paper.

Funding: This research was supported by the Keimyung University Research Grant of 2019.

Conflicts of Interest: The authors declare no conflicts of interest.

References

1. Balandin, A.A. Thermal properties of graphene and nanostructured carbon materials. *Nat. Mater.* **2011**, *10*, 569–581. [CrossRef] [PubMed]
2. Inagaki, M.; Kaburagi, Y.; Hishiyama, Y. Thermal Management Material: Graphite. *Adv. Eng. Mater.* **2014**, *16*, 494–506. [CrossRef]
3. Jiang, B.; Wang, H.; Wen, G.; Wang, E.; Fang, X.; Liu, G.; Zhou, W. Copper–graphite–copper sandwich: Superior heat spreader with excellent heat-dissipation ability and good weldability. *RSC Adv.* **2016**, *6*, 25128–25136. [CrossRef]
4. Yan, Z.; Liu, G.; Khan, J.M.; Balandin, A.A. Graphene quilts for thermal management of high-power GaN transistors. *Nat. Commun.* **2012**, *3*, 827. [CrossRef] [PubMed]
5. Park, C.G.; Lee, T.H.; Lee, T.K.; Jeong, M.Y. A Study on the Optimization of Heat Dissipation in Flip-chip Package. *J. Microelectron. Packag. Soc.* **2013**, *20*, 75–80. [CrossRef]
6. Ko, Y.-H.; Choi, K.; Kim, S.W.; Yu, D.-Y.; Bang, J.; Kim, T.-S. Trends of Researches and Technologies of Electronic Packaging Using Graphene. *J. Microelectron. Packag. Soc.* **2016**, *23*, 1–10. [CrossRef]
7. Abyzov, A.M.; Kidalov, S.V.; Shakhov, F.M. High thermal conductivity composite of diamond particles with tungsten coating in a copper matrix for heat sink application. *Appl. Therm. Eng.* **2012**, *48*, 72–80. [CrossRef]
8. Boden, A.; Boerner, B.; Kusch, P.; Firkowska, I.; Reich, S. Nanoplatelet Size to Control the Alignment and Thermal Conductivity in Copper–Graphite Composites. *Nano Lett.* **2014**, *14*, 3640–3644. [CrossRef]
9. Firkowska, I.; Boden, A.; Boerner, B.; Reich, S. The Origin of High Thermal Conductivity and Ultralow Thermal Expansion in Copper–Graphite Composites. *Nano Lett.* **2015**, *15*, 4745–4751. [CrossRef]
10. Slack, G.A. Thermal Conductivity of Pure and Impure Silicon, Silicon Carbide, and Diamond. *J. Appl. Phys.* **1964**, *35*, 3460. [CrossRef]
11. Liu, Z.; Guo, Q.; Shi, J.; Zhai, G.; Liu, L. Graphite blocks with high thermal conductivity derived from natural graphite flake. *Carbon* **2008**, *46*, 414–421. [CrossRef]
12. Hui, C.; Zhang, Y.; Zhang, L.; Sun, R.; Liu, F. Crumpling of a pyrolytic graphite sheet. *J. Appl. Phys.* **2013**, *114*, 163512. [CrossRef]
13. Wen, C.-Y.; Huang, G.-W. Application of a thermally conductive pyrolytic graphite sheet to thermal management of a PEM fuel cell. *J. Power Sources* **2008**, *178*, 132–140. [CrossRef]
14. Kim, P.; Shi, L.; Majumdar, A.; McEuen, P.L. Thermal Transport Measurements of Individual Multiwalled Nanotubes. *Phys. Rev. Lett.* **2001**, *87*, 215502. [CrossRef] [PubMed]
15. Balandin, A.A.; Ghosh, S.; Bao, W.; Calizo, I.; Teweldebrhan, D.; Miao, F.; Lau, C.N. Superior Thermal Conductivity of Single-Layer Graphene. *Nano Lett.* **2008**, *8*, 902–907. [CrossRef]
16. Seol, J.H.; Jo, I.; Moore, A.L.; Lindsay, L.; Aitken, Z.H.; Pettes, M.T.; Li, X.; Yao, Z.; Huang, R.; Broido, D.; et al. Two-Dimensional Phonon Transport in Supported Graphene. *Science* **2010**, *328*, 213–216. [CrossRef]

17. Song, W.-L.; Veca, L.M.; Kong, C.Y.; Ghose, S.; Connell, J.W.; Wang, P.; Cao, L.; Lin, Y.; Meziani, M.J.; Qian, H.; et al. Polymeric nanocomposites with graphene sheets—Materials and device for superior thermal transport properties. *Polymer* **2012**, *53*, 3910–3916. [CrossRef]

18. Xiang, C.; Young, C.C.; Hwang, C.-C.; Cerioti, G.; Tour, J.M.; Wang, X.; Yan, Z.; Hwang, C.; Lin, J.; Kono, J.; et al. Large Flake Graphene Oxide Fibers with Unconventional 100% Knot Efficiency and Highly Aligned Small Flake Graphene Oxide Fibers. *Adv. Mater.* **2013**, *25*, 4592–4597. [CrossRef]

19. Gwinn, J.; Webb, R. Performance and testing of thermal interface materials. *Microelectron. J.* **2003**, *34*, 215–222. [CrossRef]

20. Yang, W.; Zhou, L.; Peng, K.; Zhu, J.; Wan, L. Effect of tungsten addition on thermal conductivity of graphite/copper composites. *Compos. Part B Eng.* **2013**, *55*, 1–4. [CrossRef]

21. Kang, Q.; He, X.; Ren, S.; Zhang, L.; Wu, M.; Liu, T.; Liu, Q.; Guo, C.; Qu, X. Preparation of high thermal conductivity copper–diamond composites using molybdenum carbide-coated diamond particles. *J. Mater. Sci.* **2013**, *48*, 6133–6140. [CrossRef]

22. Prieto, R.; Molina, J.; Narciso, J.; Louis, E. Fabrication and properties of graphite flakes/metal composites for thermal management applications. *Scr. Mater.* **2008**, *59*, 11–14. [CrossRef]

23. Jagannadham, K. Orientation dependence of thermal conductivity in copper-graphene composites. *J. Appl. Phys.* **2011**, *110*, 74901. [CrossRef]

24. Jagannadham, K. Thermal Conductivity of Copper-Graphene Composite Films Synthesized by Electrochemical Deposition with Exfoliated Graphene Platelets. *Metall. Mater. Trans. B* **2012**, *43*, 316–324. [CrossRef]

25. Liu, Q.; He, X.-B.; Ren, S.-B.; Zhang, C.; Ting-Ting, L.; Qu, X.-H. Thermophysical properties and microstructure of graphite flake/copper composites processed by electroless copper coating. *J. Alloys Compd.* **2014**, *587*, 255–259. [CrossRef]

26. Goli, P.; Ning, H.; Li, X.; Lu, C.Y.; Novoselov, K.S.; Balandin, A.A. Thermal Properties of Graphene–Copper–Graphene Heterogeneous Films. *Nano Lett.* **2014**, *14*, 1497–1503. [CrossRef] [PubMed]

27. Zhang, D.-D.; Zhan, Z.-J. Experimental investigation of interfaces in graphene materials/copper composites from a new perspective. *RSC Adv.* **2016**, *6*, 52219–52226. [CrossRef]

28. Wang, L.; Yang, Z.; Cui, Y.; Wei, B.; Xu, S.; Sheng, J.; Wang, M.; Zhu, Y.; Fei, W. Graphene-copper composite with micro-layered grains and ultrahigh strength. *Sci. Rep.* **2017**, *7*, 41896. [CrossRef]

29. Hwang, J.; Yoon, T.; Jin, S.H.; Lee, J.; Kim, T.-S.; Hong, S.H.; Jeon, S. Enhanced Mechanical Properties of Graphene/Copper Nanocomposites Using a Molecular-Level Mixing Process. *Adv. Mater.* **2013**, *25*, 6724–6729. [CrossRef]

30. Wang, L.; Cui, Y.; Li, B.; Yang, S.; Liu, Z.; Vajtai, R.; Fei, W. High apparent strengthening efficiency for reduced graphene oxide in copper matrix composites produced by molecule-lever mixing and high-shear mixing. *RSC Adv.* **2015**, *5*, 51193–51200. [CrossRef]

31. Choi, I.; Jeong, H.Y.; Jung, D.Y.; Byun, M.; Choi, C.-G.; Hong, B.H.; Choi, S.-Y.; Lee, K.J. Laser-Induced Solid-Phase Doped Graphene. *ACS Nano* **2014**, *8*, 7671–7677. [CrossRef]

32. Choi, I.; Jeong, H.Y.; Shin, H.; Kang, G.; Byun, M.; Kim, H.; Chitu, A.M.; Im, J.S.; Ruoff, R.S.; Choi, S.-Y.; et al. Laser-induced phase separation of silicon carbide. *Nat. Commun.* **2016**, *7*, 13562. [CrossRef] [PubMed]

33. Stankovich, S.; Dikin, D.A.; Piner, R.D.; Kohlhaas, K.A.; Kleinhammes, A.; Jia, Y.; Wu, Y.; Nguyen, S.T.; Ruoff, R.S. Synthesis of graphene-based nanosheets via chemical reduction of exfoliated graphite oxide. *Carbon* **2007**, *45*, 1558–1565. [CrossRef]

34. Liu, X.-W.; Mao, J.-J.; Liu, P.-D.; Wei, X.-W. Fabrication of metal-graphene hybrid materials by electroless deposition. *Carbon* **2011**, *49*, 477–483. [CrossRef]

35. Sato, K.; Saito, R.; Oyama, Y.; Jiang, J.; Cançado, L.; Pimenta, M.; Jorio, A.; Samsonidze, G.; Dresselhaus, G.; Dresselhaus, M.; et al. D-band Raman intensity of graphitic materials as a function of laser energy and crystallite size. *Chem. Phys. Lett.* **2006**, *427*, 117–121. [CrossRef]

36. Barcena, J.; Maudes, J.; Coleto, J.; Baldonedo, J.L.; De Salazar, J.M.G. Microstructural study of vapour grown carbon nanofibre/copper composites. *Compos. Sci. Technol.* **2008**, *68*, 1384–1391. [CrossRef]

37. Mai, Y.; Zhang, D.; Qiao, Y.; Gu, C.; Wang, X.; Tu, J. MnO/reduced graphene oxide sheet hybrid as an anode for Li-ion batteries with enhanced lithium storage performance. *J. Power Sources* **2012**, *216*, 201–207. [CrossRef]

38. Song, H.; Wang, B.; Zhou, Q.; Xiao, J.; Jia, X. Preparation and tribological properties of MoS2/graphene oxide composites. *Appl. Surf. Sci.* **2017**, *419*, 24–34. [CrossRef]
39. Schmidt, A.J.; Collins, K.C.; Minnich, A.J.; Chen, G. Thermal conductance and phonon transmissivity of metal–graphite interfaces. *J. Appl. Phys.* **2010**, *107*, 104907. [CrossRef]

Article

Effect of Deep Cryogenic Treatment on Microstructure and Properties of Sintered Fe–Co–Cu-Based Diamond Composites

Siqi Li [1,2], Wenhao Dai [1,2], Zhe Han [1,2], Xinzhe Zhao [1,2] and Baochang Liu [1,2,3,*]

1   College of Construction Engineering, Jilin University, Changchun 130026, China
2   Key Laboratory of Drilling and Exploitation Technology in Complex Conditions of Ministry of Natural Resources, Changchun 130026, China
3   State Key Laboratory of Superhard Materials, Changchun 130021, China
*   Correspondence: liubc@jlu.edu.cn; Tel.: +86-431-8850-2891

Received: 15 July 2019; Accepted: 12 August 2019; Published: 15 August 2019

Abstract: Metal-matrix-impregnated diamond composites are used for fabricating many kinds of diamond tools. In the efforts to satisfy the increasing engineering requirements, researchers have brought more attention to find novel methods of enhancing the performance of impregnated diamond composites. In this study, deep cryogenic treatment was applied to Fe–Co–Cu-based diamond composites to improve their performance. Relative density, hardness, bending strength, and grinding ratio of matrix and diamond composite samples were measured by a series of tests. Meanwhile, the fracture morphologies of all samples after the bending strength test were observed and analyzed by scanning electron microscope. The results showed that the hardness and bending strength of matrix increased slightly after deep cryogenic treatment. The grinding ratio of impregnated diamond composites exhibited a great increase by 32.9% as a result of deep cryogenic treatment. The strengthening mechanism was analyzed in detail. The Fe–Co–Cu-based impregnated composites subjected to deep cryogenic treatment for 1 h exhibited the best overall performance.

Keywords: deep cryogenic treatment; impregnated diamond composites; metal matrix; wear resistance

---

1. Introduction

Metal-matrix-impregnated diamond composites are seeing wide use in the fabrication of diamond tools for cutting, grinding, polishing, and drilling hard and brittle materials such as ceramic, rock, and concrete [1–7]. The metal-matrix-impregnated diamond composites prepared by powder metallurgy consist of two parts: diamond grits and metal matrix. The diamond grits embedded in the metal matrix are the working element, and the metal matrix has a decisive effect on whether or not the diamond grits can function fully and effectively [2,8].

At present, pre-alloyed powders have successfully realized industrial application in diamond tools, including diamond cutting segments and diamond drill bits [9–12]. The component uniformity is one of the advantages of pre-alloyed powders, because each alloying particle includes the entire composition, which means the segregation can be avoided [13]. In addition, the pre-alloying of powders can decrease the activation energy, which is conductive to reducing the sintering temperature, which in turn can avoid diamond graphitization [14]. The Fe–Co–Cu matrix, sintered from pre-alloyed Fe–Co–Cu powders, is usually used as the matrix in diamond tool manufacturing [4,10]. It has high strength and adjustable properties suitable for different kinds of hard materials, ensuring that the diamond grits contact the hard materials effectively, maintaining an abrasive cutting surface. Nevertheless, the hard and complex service conditions such as elevated temperature, high impact stresses, and hydro-abrasion demand that the mechanical properties and wear resistance be further improved [7,15].

Deep cryogenic treatment (DCT) is a supplement of conventional heat treatment used to modify materials properties in order to enhance their performance [16,17]. Recently, some researchers found that deep cryogenic treatment can improve the hardness, compression strength, fatigue resistance, and wear resistance of tool steels [18–23]. The transformation of retained austenite to martensite, the release of residual stress, and the precipitation of ultra-fine carbide particles are the main mechanisms of cryogenic treatment in tool steels [24]. Moreover, some studies have indicated that deep cryogenic treatment can increase the wear resistance and service life of polycrystalline diamond compact (PDC), which is ascribed to the change of stress state [25]. It can be predicted that deep cryogenic treatment is a promising way to improve the properties of metal-matrix-impregnated diamond composites. However, the effect of deep cryogenic treatment on the microstructure, mechanical properties, and wear resistance of metal-matrix-impregnated diamond composites has been rarely studied.

In this study, deep cryogenic treatment was applied to Fe–Co–Cu-based diamond composites. The effects of deep cryogenic treatment on the microstructure, mechanical properties, and wear resistance of Fe–Co–Cu-based diamond composites were investigated and discussed. This work is aimed at seeking the optimal treatment time to meet the requirements of mechanical properties and wear resistance.

2. Materials and Methods

2.1. Sample Preparation

Two kinds of samples, including matrix samples with and without impregnated diamond grits, were prepared for the property tests. The composition of the Fe–Co–Cu pre-alloy powder (99.9% purity, average particle size of 15 µm, Qinhuangdao, China) used in this work is given in Table 1. The diamond grits (20 vol% concentration, synthetic, 270–380 µm, Zhengzhou, China) were added into the Fe–Co–Cu pre-alloy powder for preparing diamond composite samples through a three-axle mixer for 4 h. The samples (size: $38 \times 8 \times 5$ mm^3) were sintered by hot-pressing in graphite molds at 840 °C and a pressure of 20 MPa for 4 min. Sintering temperature, pressure, and time were selected on the basis of previous research and published literature [4,9,10]. The composition, deep cryogenic treatment process, and mechanical properties of samples are given in Table 2.

Table 1. The composition of the Fe–Co–Cu pre-alloy powder used in this work.

Composition	Fe	Co	Cu
Content (wt%)	54	32	14

Table 2. The composition, deep cryogenic treatment (DCT) process, and mechanical properties of samples.

Samples	Composition	DCT Time (h)	Relative Density (%)	Hardness (HRC)	Bending Strength (MPa)
S0	Matrix	-	97.3	38.1 ± 0.6	1620 ± 25
S1	Matrix	1	97.4	39.1 ± 0.5	1639 ± 23
S2	Matrix	2	97.3	38.8 ± 0.6	1720 ± 29
S3	Matrix	3	97.5	39.1 ± 0.3	1747 ± 27
SD0	Matrix + diamond	-	95.7	-	740 ± 35
SD1	Matrix + diamond	1	96.7	-	796 ± 33
SD2	Matrix + diamond	2	96.2	-	752 ± 30
SD3	Matrix + diamond	3	96.4	-	761 ± 50

2.2. Deep Cryogenic Treatment

A program-controlled cryogenic system (CDW-196, Jinan, China) was used for the deep cryogenic treatment. The schematic diagram is shown in Figure 1. In this device, liquid nitrogen acting as the coolant flows into the distributor to be uniformly dispersed, and then evaporates into nitrogen. Under the action of fan rotation, the gas flow enters the cryogenic tank and exchanges heat with samples. Then, the relatively hot nitrogen gas is emitted into the atmosphere from the exhaust port. Parameters including temperature, cooling rate, and soaking time can be accurately controlled by the computer. In this experiment, the samples were cooled from room temperature (25 °C) down to −190 °C at the cooling rate of 5 °C/min, holding for 1, 2, and 3 h, respectively, and then warmed up to room temperature at a speed of 1 °C/min.

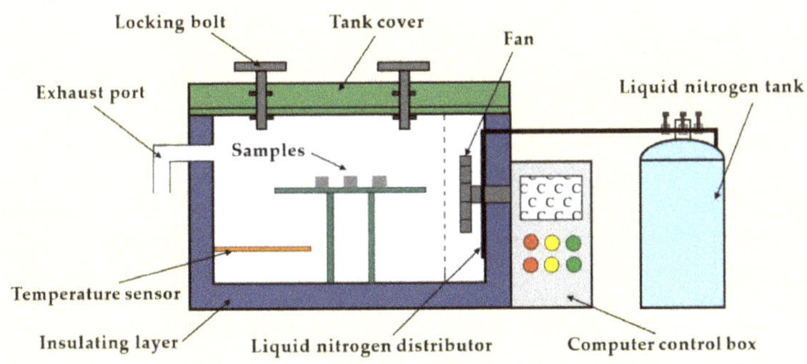

Figure 1. The schematic diagram of CDW-196 cryogenic treatment system.

2.3. Characterization

The microstructure of composites was characterized by scanning electron microscopy (SEM, Hitachi S-4800, Tokyo, Japan). The phase analysis was evaluated by X-ray diffractometry (TTR-III, Tokyo, Japan). A high-precision density tester (Dahometer, DE-120M, Hangzhou, China) was used to measure the density of samples via the Archimedes method. A Rockwell hardness tester (Huayin HRS-150, Yantai, China) was used to measure the Rockwell hardness scale C (HRC) of samples. Bending strength for each sample was measured by using a three-point bending test machine (DDL 100, Changchun, China), and was calculated by following formula:

$$\sigma = 3PL/2bh^2, \tag{1}$$

where σ is the bending strength, MPa; P is the axial load, N; L is the length of the sample holder, 24 mm; b is the width, 5 mm; and h is the height, 8 mm.

A grinding ratio measurement device (DHM-2, Changchun, China) was applied to measure the grinding ratio. A schematic diagram of the grinding ratio test is illustrated in Figure 2. SiC grinding wheels with diameter of 100 mm and thickness of 20 mm were used as wear counterparts. The testing parameters involved were: linear velocity 15 m/s, swing frequency 30 min^{-1}, load 500 g, and grinding time ≥100 s. The grinding ratio Ra was calculated by the formula:

$$Ra = \Delta Mg/\Delta Ms, \tag{2}$$

where ΔMg is the weight loss of the SiC grinding wheel, and ΔMs is the weight loss of sample.

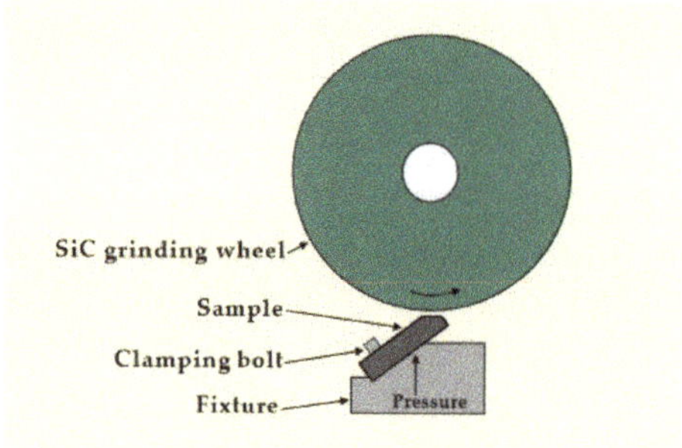

Figure 2. Schematic diagram of the grinding ratio test.

3. Results and Discussion

3.1. Composition and Microstructures

The XRD patterns of matrix samples with different deep cryogenic treatment times are shown in Figure 3. The elemental phase Fe, Cu, and intermetallic phases of Co_3Fe_7, FeCo, and $FeCu_4$ were observed in all samples. Note that the deep cryogenic treatment did not lead to an obvious change in the lattice parameters of phases, as determined on the basis of the XRD analysis.

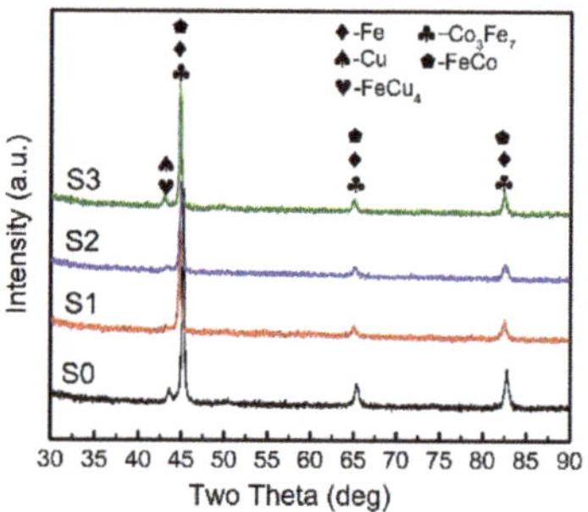

Figure 3. XRD patterns of matrix samples.

SEM micrographs for the bending fracture surface of matrix samples S0 and S3 are presented in Figure 4. The size of the grains was uniform and there was no abnormal grain growth. Examinations of the fracture surfaces indicated that the fracture mechanisms of matrix samples were trans-crystalline and inter-crystalline fracture, with inter-crystalline fracture being more common. It is important to note that the trans-crystalline fracture phenomenon of sample S3 (DCT for 3 h) was more obvious than in S0, which can be attributed to the development of stresses. The matrix material contained elemental phase Fe, Cu, and intermetallic phases of Co_3Fe_7, FeCo, and $FeCu_4$, which have different coefficients of thermal expansion (CTEs). During the deep cryogenic treatment, the high-CTE phase was subjected to compressive stresses and the low-CTE phase to tensile ones. These uneven structural changes that occurred because of normal contractions were opposed by transformation expansion at this temperature [26]. This phenomenon developed residual stresses in the matrix material, which led to an increase in dislocation density. When external stress was applied, the difficulty of grain dislocation motion increased, causing the grain to break and more easily spread inside the grain to produce trans-crystalline fracture.

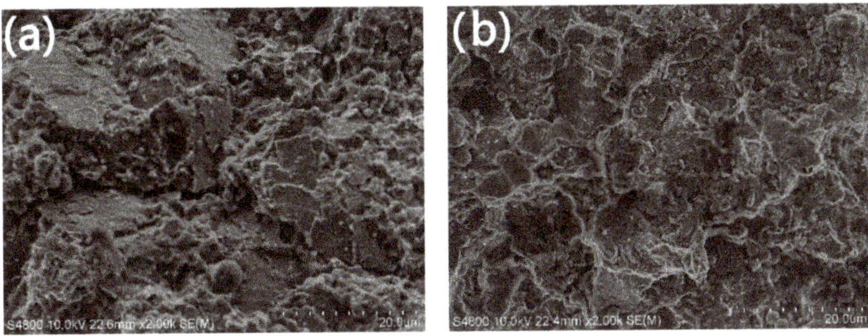

Figure 4. The fracture morphologies of samples (**a**) S0 and (**b**) S3.

SEM micrographs for the bending fracture surface of impregnated diamond samples SD0 and SD1 are given in Figure 5. The figure reveals that diamond grits were embedded in the metal matrix. As marked in Figure 5a,d, the average width of the interface crack between diamond grits and matrix decreased from 2.8 to 2.1 μm after deep cryogenic treatment for 1 h. The change of crack width can be attributed to the difference of thermal conductivity and thermal expansion coefficient of diamond and matrix materials. Table 3 shows the specific values of thermal conductivity and coefficient of thermal expansion [27]. During the deep cryogenic treatment process, the stress state changed significantly, and the matrix was subjected to tensile stress. At a certain moment, the tensile stress exceeded the micro-yield strength of the matrix, which led to the plastic deformation of the matrix materials. The smaller crack width indicates that the diamond grits were held more firmly by the matrix, which is conducive to the mechanical properties and wear resistance of impregnated diamond composites. The average crack widths of SD2 (DCT for 2 h) and SD3 (DCT for 3 h) were 2.6 and 2.5 μm, respectively. Cracks in the diamond/matrix interface have a negative impact on relative density, which is consistent with the test results in Table 2.

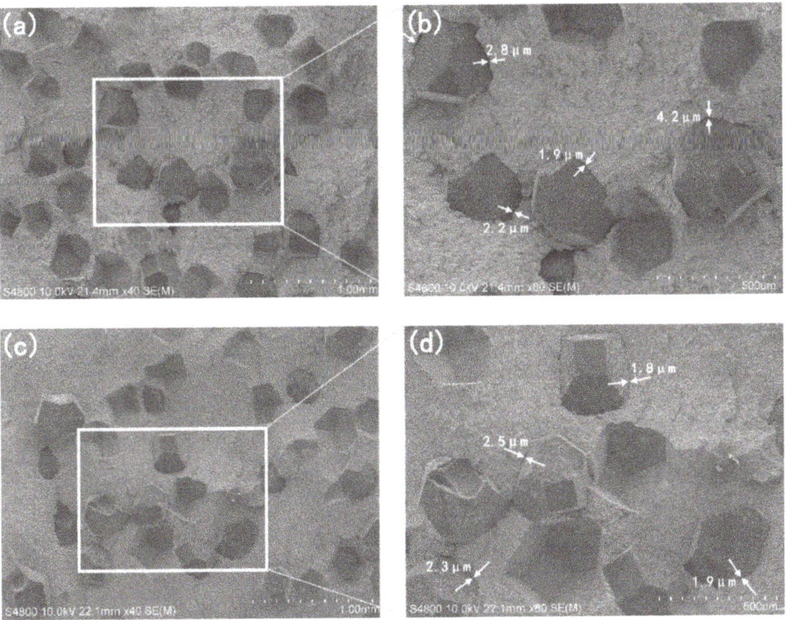

Figure 5. The fracture morphologies of impregnated samples: (**a**,**b**) SD0; (**c**,**d**) SD1. The crack widths were measured by image processing software at high magnification.

Table 3. The values of thermal conductivity and coefficient of thermal expansion.

Composition	Thermal Conductivity (W/m·K)	Thermal Expansion Coefficient (K^{-1})
Diamond	2000	$\approx 1 \times 10^{-6}$
Matrix material	125	$\geq 13 \times 10^{-6}$

3.2. Mechanical Properties

The relative density of all tested samples is shown in Figure 6a. It can be seen that deep cryogenic treatment had little effect on the relative density of matrix samples. For impregnated samples, the relative density increased after deep cryogenic treatment due to the decrease of crack width between diamond and matrix.

As displayed in Figure 6b,c the hardness and bending strength of matrix samples increased slightly after deep cryogenic treatment. The results can be primarily related to the development of stresses. The development of stresses was due to the difference in thermal expansion coefficients between different phases. The development of stresses can increase the dislocation density, which leads to the improvement of the hardness and bending strength.

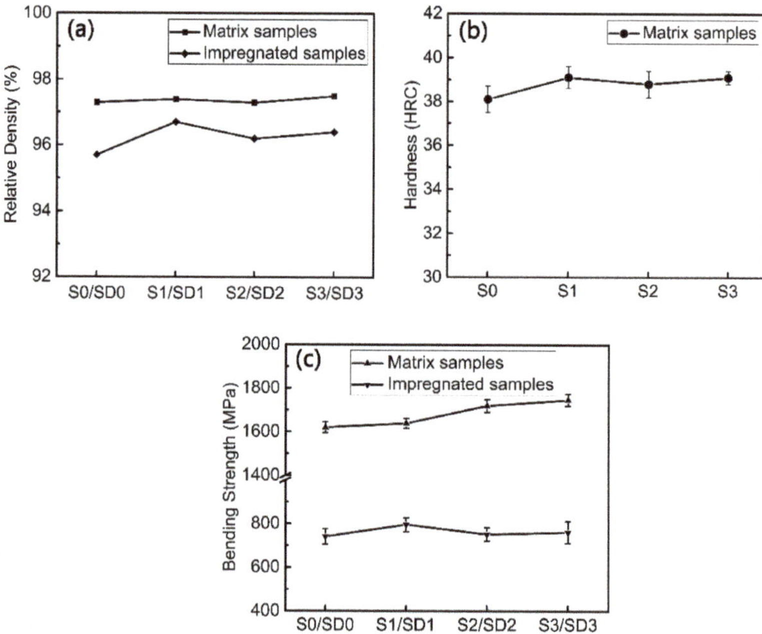

Figure 6. Mechanical properties of samples. (**a**) The relative density of all samples; (**b**) The HRC values of matrix samples; (**c**) The bending strength of all samples.

The bending strength of all samples is illustrated in Figure 6c. Sample SD1 showed the highest value and the values of SD2 and SD3 were similar to SD0. The interface structure in Figure 5 observed between diamond grits and matrix of all impregnated samples is concordant with the bending strength test results. The average crack width decreased from 2.8 to 2.1 μm after deep cryogenic treatment for 1 h. This structure feature means that the diamond grits were held by the matrix more firmly, which benefits the stress transfer, resulting in a high bending strength.

3.3. Wear Resistance

The grinding ratio is an important indicator to evaluate the wear resistance and tribological performance of diamond-impregnated composites [28,29]. The bar chart in Figure 7 summarizes the results for the grinding ratio of impregnated diamond samples with different deep cryogenic treatment times. The grinding ratio increased remarkably after deep cryogenic treatment, proving that deep cryogenic treatment improved the wear resistance of the impregnated diamond composites. With increasing deep cryogenic treatment time, the grinding ratio first increased and then decreased. The grinding ratio of sample SD1 showed a 32.9% increase in comparison with SD0.

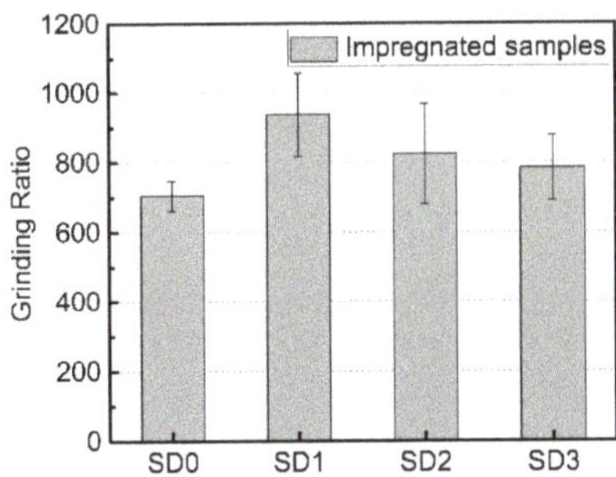

Figure 7. Results of the grinding ratio test of impregnated diamond samples.

The grinding ratio of impregnated samples depends largely on the bending strength. Some related studies analyzed the theoretical relationship between bending strength and wear resistance [5,30]. A lower bending strength means that the metal matrix supports the diamond grits weakly, leading to the pull-out of diamond grits from the matrix under complex stress conditions during working. Hence, the grinding ratio of SD1 subjected to 1 h of deep cryogenic treatment was higher than the reference one (SD0) without deep cryogenic treatment.

The grinding ratios of SD2 and SD3 were larger than that of SD0. However, the bending strengths of SD2 and SD3 were similar to that of SD0, indicating that there are other factors affecting the wear resistance. The mismatch in the coefficient of thermal expansion between diamond ($\approx 1 \times 10^{-6}$ K^{-1}) and matrix material ($\geq 13 \times 10^{-6}$ K^{-1}) needs to be given further consideration. During the deep cryogenic treatment process, the metal-matrix-impregnated diamond composites were subjected to a thermal shock, with a temperature drop of 215 °C, which increased the residual compressive stress on diamond grits. This change of the stress state increased the diamond retention capacity of the matrix. Furthermore, the matrix materials penetrated into the microcracks on the surface of diamond grits, which could also improve the diamond retention capacity of the matrix. Facing the action of complex alternating cutting force, diamond grits usually exhibit a tendency to rotate first and then pull out of the matrix easily. The improvement of the matrix's diamond retention capacity can reduce the rotating tendency, which is conducive to improving the wear resistance of diamond composites.

4. Conclusions

The effects of deep cryogenic treatment on the microstructure, mechanical properties, and wear resistance of Fe–Co–Cu-based diamond-impregnated composites were investigated. The experimental results indicated that the hardness and bending strength of matrix increased slightly after deep cryogenic treatment. The grinding ratio of the impregnated diamond composites increased remarkably by 32.9% as a result of deep cryogenic treatment. The strengthening mechanism was analyzed in detail. The conjoint effects of bending strength and diamond retention capability on the wear resistance of the diamond composites was revealed. The Fe–Co–Cu-based impregnated composites subjected to deep cryogenic treatment for 1 h exhibited the optimal overall performance, thus providing guidance for further scientific research and actual applications.

Author Contributions: B.L. and S.L. conceived and designed the experiments; S.L. and W.D. performed the experiments; Z.H. and X.Z. analyzed the data; B.L. contributed analysis tools; S.L. wrote the paper.

Funding: This work was financially supported by the National Natural Science Foundation of China (41572357) and the Science and Technology project of Jilin Province Education Department (JJKH20180087KJ).

Conflicts of Interest: The authors declare no conflicts of interest.

References

1. Tönshoff, H.K.; Hillmann-Apmann, H.; Asche, J. Diamond tools in stone and civil engineering industry cutting principles, wear and applications. *Diam. Relat. Mater.* **2002**, *250*, 103–109. [CrossRef]
2. Zhao, X.; Duan, L. A Review of the Diamond Retention Capacity of Metal Bond Matrices. *Metals* **2018**, *8*, 307. [CrossRef]
3. Boland, J.N.; Li, X.S. Microstructural Characterisation and Wear Behaviour of Diamond Composite Materials. *Materials* **2010**, *3*, 1390–1419. [CrossRef]
4. Li, W.; Zhan, J.; Wang, S.; Dong, H.; Li, Y.; Liu, Y. Characterizations and mechanical properties of impregnated diamond segment using Cu-Fe-Co metal matrix. *Rare Metals* **2012**, *31*, 81–87. [CrossRef]
5. Di Ilio, A.; Togna, A. A theoretical wear model for diamond tools in stone cutting. *Int. J. Mach. Tools Manuf.* **2003**, *43*, 1171–1177. [CrossRef]
6. Sun, Y.; Wu, H.; Li, M.; Meng, Q.; Gao, K.; Lu, X.; Liu, B. The Effect of ZrO2 Nanoparticles on the Microstructure and Properties of Sintered WC-Bronze-Based Diamond Composites. *Materials* **2016**, *9*, 343. [CrossRef] [PubMed]
7. Li, S.; Han, Z.; Meng, Q.; Zhao, X.; Cao, X.; Liu, B. Effect of WC Nanoparticles on the Microstructure and Properties of WC-Bronze-Ni-Mn Based Diamond Composites. *Appl. Sci.* **2018**, *8*, 1501. [CrossRef]
8. Artini, C.; Muolo, M.L.; Passerone, A. Diamond–metal interfaces in cutting tools: A review. *J. Mater. Sci.* **2011**, *47*, 3252–3264. [CrossRef]
9. Xie, D.; Wan, L.; Song, D.; Wang, S.; Lin, F.; Pan, X.; Xu, J. Pressureless sintering curve and sintering activation energy of Fe–Co–Cu pre-alloyed powders. *Mater. Des.* **2015**, *87*, 482–487. [CrossRef]
10. Xie, D.; Xiao, L.; Lin, F.; Pan, X.; Su, Y.; Fang, X.; Qi, H.; Chen, C. Thermal Analysis of FeCoCu Pre-Alloyed Powders Used for Diamond Tools. *J. Superhard Mater.* **2018**, *40*, 110–117. [CrossRef]
11. Dai, H.; Wang, L.; Zhang, J.; Liu, Y.; Wang, Y.; Wang, L.; Wan, X. Iron based partially pre-alloyed powders as matrix materials for diamond tools. *Powder Metall.* **2015**, *58*, 83–86. [CrossRef]
12. Wang, F.; Wakoh, K.; Li, Y.; Ito, S.; Yamanaka, K.; Koizumi, Y.; Chiba, A. Study of microstructure evolution and properties of Cu-Fe microcomposites produced by a pre-alloyed powder method. *Mater. Des.* **2017**, *126*, 64–72. [CrossRef]
13. Chu, Z.-Q.; Guo, X.-Y.; Liu, D.-H.; Tan, Y.-X.; Li, D.; Tian, Q.-H. Application of pre-alloyed powders for diamond tools by ultrahigh pressure water atomization. *Trans. Nonferr. Metals Soc. China* **2016**, *26*, 2665–2671. [CrossRef]
14. Zhao, X.; Li, J.; Duan, L.; Tan, S.; Fang, X. Effect of Fe-based pre-alloyed powder on the microstructure and holding strength of impregnated diamond bit matrix. *Int. J. Refract. Metals Hard Mater.* **2019**, *79*, 115–122. [CrossRef]
15. Levashov, E.; Kurbatkina, V.; Alexandr, Z. Improved Mechanical and Tribological Properties of Metal-Matrix Composites Dispersion-Strengthened by Nanoparticles. *Materials* **2009**, *3*, 97–109. [CrossRef]
16. Gu, K.; Zhang, H.; Zhao, B.; Wang, J.; Zhou, Y.; Li, Z. Effect of cryogenic treatment and aging treatment on the tensile properties and microstructure of Ti–6Al–4V alloy. *Mater. Sci. Eng. A* **2013**, *584*, 170–176. [CrossRef]
17. Gao, Y.; Luo, B.-H.; Bai, Z.-h.; Zhu, B.; Ouyang, S. Effects of deep cryogenic treatment on the microstructure and properties of WC Fe Ni cemented carbides. *Int. J. Refract. Metals Hard Mater.* **2016**, *58*, 42–50. [CrossRef]
18. Li, H.; Tong, W.; Cui, J.; Zhang, H.; Chen, L.; Zuo, L. The influence of deep cryogenic treatment on the properties of high-vanadium alloy steel. *Mater. Sci. Eng. A* **2016**, *662*, 356–362. [CrossRef]
19. Gill, S.S.; Singh, H.; Singh, R.; Singh, J. Cryoprocessing of cutting tool materials—A review. *Int. J. Adv. Manuf. Technol.* **2009**, *48*, 175–192. [CrossRef]

20. Jimbert, P.; Iturrondobeitia, M.; Ibarretxe, J.; Fernandez-Martinez, R. Influence of Cryogenic Treatment on Wear Resistance and Microstructure of AISI A8 Tool Steel. *Metals* **2018**, *8*, 1038. [CrossRef]

21. Gonçalves, V.R.M.; Podgornik, B.; Leskovšek, V.; Totten, G.E.; Canale, L.d.C.F. Influence of Deep Cryogenic Treatment on the Mechanical Properties of Spring Steels. *J. Mater. Eng. Perform.* **2019**, *28*, 769–775.

22. Nguyen, L.T.H.; Hwang, J.-S.; Kim, M.-S.; Kim, J.-H.; Kim, S.-K.; Lee, J.-M. Charpy Impact Properties of Hydrogen-Exposed 316L Stainless Steel at Ambient and Cryogenic Temperatures. *Metals* **2019**, *9*, 625. [CrossRef]

23. Li, B.; Li, C.; Wang, Y.; Jin, X. Effect of Cryogenic Treatment on Microstructure and Wear Resistance of Carburized 20CrNi2MoV Steel. *Metals* **2018**, *8*, 808. [CrossRef]

24. Candane, D. Effect of Cryogenic Treatment on Microstructure and Wear Characteristics of AISI M35 HSS. *Int. J. Mater. Sci. Appl.* **2013**, *2*, 56. [CrossRef]

25. Khaidarov, K.K.; Kozhogulov, O.C.; Makarov, V.P. Strength index and structure of PCD after cryogenic treatment. *Strength Mater.* **1996**, *28*, 37–40. [CrossRef]

26. Kalsi, N.S.; Sehgal, R.; Sharma, V.S. Effect of tempering after cryogenic treatment of tungsten carbide–cobalt bonded insert. *Bull. Mater. Sci.* **2014**, *37*, 327–335. [CrossRef]

27. Sun, L.B.; Sun, Y.F. *Metal Materials Physical Properties Handbook*; Mechanical Industry Press: Beijing, China, 2011.

28. Wang, Z.; Zhang, Z.; Sun, Y.; Gao, K.; Liang, Y.; Li, X.; Ren, L. Wear behavior of bionic impregnated diamond bits. *Tribol. Int.* **2016**, *94*, 217–222. [CrossRef]

29. Li, M.; Sun, Y.; Meng, Q.; Wu, H.; Gao, K.; Liu, B. Fabrication of Fe-Based Diamond Composites by Pressureless Infiltration. *Materials* **2016**, *9*, 1006. [CrossRef]

30. Janusz, K. Theoretical analysis of stone sawing with diamonds. *J. Mater. Process. Technol.* **2002**, *123*, 146–154.

Article

Effect of 0.05 wt.% Pr Addition on Microstructure and Shear Strength of Sn-0.3Ag-0.7Cu/Cu Solder Joint during the Thermal Aging Process

Jie Wu [1,2], Songbai Xue [1,*], Jingwen Wang [1] and Guoqiang Huang [1]

[1]   College of Materials Science and Technology, Nanjing University of Aeronautics and Astronautics, Nanjing 210016, China
[2]   Department of Mechanical, Aerospace and Biomedical Engineering, University of Tennessee Knoxville, Knoxville, TN 37996, USA
*   Correspondence: xuesb@nuaa.edu.cn

Received: 1 August 2019; Accepted: 22 August 2019; Published: 2 September 2019

Abstract: The evolution of interfacial morphology and shear strengths of the joints soldered with Sn-0.3Ag-0.7Cu (SAC0307) and SAC0307-0.05Pr aged at 150 °C for different times (h; up to 840 h) were investigated. The experiments showed the electronic joint soldered with SAC0307-0.05Pr has a much higher shear strength than that soldered with SAC0307 after each period of the aging process. This contributes to the doping of Pr atoms, "vitamins in alloys", which tend to be adsorbed on the grain surface of interfacial Cu6Sn5 IMCs, inhibiting the growth of IMCs. Theoretical analysis indicates that doping 0.05 wt.% Pr can evidently lower the growth constant of Cu6Sn5 (DCu6), while the growth constant of Cu3Sn (DCu3) decreased slightly. In addition, the electronic joint soldered with SAC0307-0.05Pr still has better ductility than that soldered with SAC0307, even after a 840-h aging process.

Keywords: SAC0307-0.05Pr solder; aging process; interfacial microstructure; fractograph

1. Introduction

In microelectronic packaging, solder joints play an electrical, mechanical, and thermal role between electronic devices and conductive substrates [1,2]. With miniaturized trends of integrated circuits in electronic devices, the reliability of solder joints becomes particularly important, especially in severe environments, such as high-temperature aging, thermal cycling, drop reliability, and so on [3–5].

In consideration of close relationship between microstructures and properties, interfacial microstructure at the solder–substrate interface should be studied to enhance the mechanical reliability of the joint. Numerous studies [6–8] have already demonstrated that one solder with more refined microstructures and thinner interfacial intermetallic compound (IMC) layers usually has a better joint reliability. To achieve this, several research groups are focused on developing composite materials doped with different types of micro- and nano-scale metallic e.g., Ag [9], Ni [10], In [11], Al [12], Co [13]; ceramic e.g., TiO2, ZrO2, Al2O3, and TiO2 etc. [9,14–16] and rare earth (RE) elements [17–19]. This is because most of these reinforcements are surface-active and prone to have adsorptions on the surfaces of the grain. Therefore, a pinning effect was exerted by them on grain growth. Sadiq et al. [20] studied the effect of RE La on the microstructures and mechanical properties of SAC alloy at a temperature of 150 °C. Corresponding results showed that the average grain size of IMCs distributed in La-doped solder was refined by up to 40% and the coarsening rate of the IMCs is slowed by 70% when compared with those in the non-doped solder. Corresponding yield stress and tensile strength are enhanced by 20% by minute lanthanum doping. In addition, the aging of nanoparticle-doped solder was also investigated, such as SAC387 solder doped with Co nanoparticle aging at 150 °C for up to 1200 h. It was found that Co nanoparticles effectively suppressed the growth of interfacial Cu_3Sn IMCs but contribute to the growth of Cu_6Sn_5. Co nanoparticles will also experience surface dissolution during reflow processes [21]. Although nanoparticles are the research hotspot presently, doping technology and agglomeration are still key issues that remain to be resolved, especially for the economically and easily available ceramic nanoparticles [22–24]. Hence, in this study, RE Pr was taken to modify microstructure and properties of solders based on previous research on solders modified by RE alloys [17,18]. For solder selection, lead-free solder has been adopted in this study according to the directives of Waste Electrical and Electronic Equipment (WEEE) and Restriction of Hazardous Substances (RoHS) [25]. Among the alternative lead-free Sn-based solder alloys such as Sn-9Zn (198 °C), Sn-58Bi (138 °C), Sn-3.5Ag (221 °C), Sn-0.7Cu (221 °C), and Sn-3.0Ag-0.5Cu (217 °C) [26–29], Sn-Ag-Cu has received wide acceptance due to its superior comprehensive properties. However, for fear of plate-like Ag_3Sn IMCs in Sn-Ag-Cu high-Ag solder ruining its drop reliability [7], the content of Ag was decreased to 0.3 wt.% in this study.

Lots of previous research has only focused on studying RE alloys on the microstructure of solder matrixes and solid–liquid reactions during soldering processes, whereas the reaction mechanism during solid–solid processes should also be given attention. This could help researchers to better understand the detailed failure mechanism of electronic device in service. Hence, this work is devoted to investigating the evolution of interfacial microstructures and the shear strength of SAC0307/Cu and SAC0307-0.05Pr/Cu solder under isothermal aging treatment (150 °C; up to 840 h). In addition, the effect of 0.05 wt.% Pr on the detailed growth kinetics of interfacial morphology at SAC0307 solder/Cu substrate interface during solid–solid reactions were also analyzed.

2. Materials and Methods

SAC0307 alloy was prepared by fusing Sn, Ag, and Cu (Purity: 99.95 wt.%) at 900 °C ± 10 °C. For fear of RE oxidation, RE Pr was doped in the Sn-Ag-Cu alloy mixture in the form of Sn-10Pr and the melting temperature was adjusted to 550 ± 1 °C. Finally, SAC0307-0.05Pr was obtained by diluting the above alloy mixture and then was shaped into bars.

Specimens for high-temperature aging were sliced from a cast sample. Interfacial specimens were made by soldering the examined solders (SAC0307; SAC0307-0.05Pr) onto Cu substrate. The aging temperature and time was selected as 150 °C and 840 h, respectively. For microstructure observation, specimens were polished with 0.3 μm Al_2O_3 suspension and then etched with the corrosion liquid (4% HNO_3-alcohol). Microstructure observation was conducted by optical microscope, Scanning electron microscope (SEM), and Energy dispersive spectrum (EDS) analysis. In addition, Image Pro-Plus software was used to calculate IMC layer's thickness.

The shear strengths of corresponding joints after each aging period were measured by a joint strength tester (Mode: STR-1000). The joint specimens made through soldering ceramic resistors (0805) on Cu substrates were put into a high-temperature environment. After shear loading, the fractographs of shear joints were studied by SEM and EDS.

3. Results and Discussion

3.1. Microstructure Evolution after Isothermal Aging

Figure 1 provided the microstructure of SAC0307, SAC0307-0.05Pr, and their aged samples (up to 840 h). As can be seen, solder doped with 0.05 wt.% Pr has a much more refined microstructure (Figure 1b) when compared with the original SAC0307 solder (Figure 1a). Obviously, IMCs in the eutectic area distributing on the Sn matrix in SAC0307-0.05Pr solder were much more refined than those in SAC0307 solder. In addition, the amount of eutectic areas was also increased after doping 0.05 wt.% Pr. After an 840-h aging treatment, IMCs in the eutectic area whether in the aged SAC0307 solder or in the aged SAC0307-0.05Pr solder grew larger (Figure 1c,d). However, the average size of the related IMCs in the aged SAC0307 were still larger in the aged SAC0307-0.05Pr and the amount of eutectic areas was also smaller than those in SAC0307-0.05Pr. It should be noted that PrSn3 phase with flower shape [30] emerged in the aged SAC0307-0.05Pr solder. Figure 1e shows the SEM image of the microstructure of aged SAC0307-0.05Pr. Also, $PrSn_3$ phase with the similar shape as that in Figure 1d can be clearly seen, which was demonstrated by the EDS analysis of Point 1, shown in Table 1. It was largely because RE Pr is a Sn-affinity element. With aging treatment, Sn and Pr atom diffusion were accelerated and finally they were reacted as $PrSn_3$. In addition, IMCs in the eutectic area was also analyzed by EDS technique and the results showed the darker ones are mainly Cu6Sn5 and the bright ones are Ag3Sn IMCs. Figure 2 gives the detailed illustration of changes in IMCs in both SAC0307 and SAC0307-0.05Pr solders during aging process. Usually, IMCs in SAC0307 grow larger with a certain rate depending on Sn, Ag, and Cu atom diffusion rates, as can be seen from Figure 2(aI–III). For SAC0307 with 0.05 wt.% Pr doped, the growth of IMCs in the eutectic area with aging process was hindered since the surface-active Pr atoms tend to adsorb on the grain surface of IMCs. Hence, the diffusion of Sn, Ag, and Cu atoms were inhibited, as shown in Figure 2(bI–III).

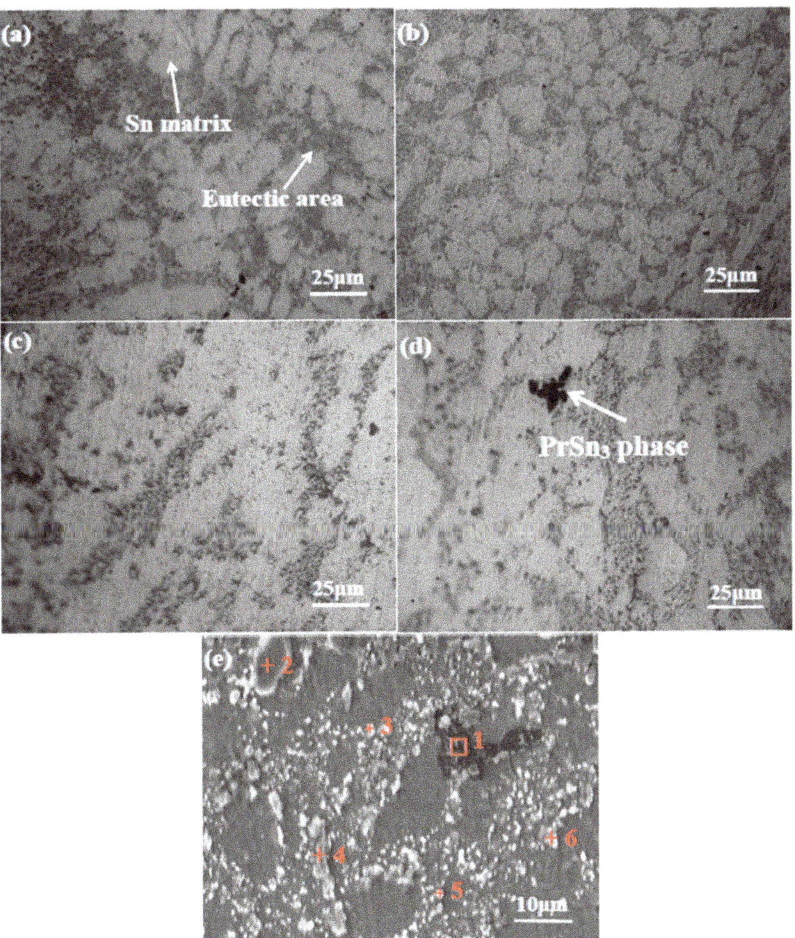

Figure 1. OM Microstructure of (**a**) SAC0307 solder; (**b**) SAC0307-0.05Pr solder; (**c**) aged SAC0307 solder at 150 °C for 840 h; (**d**) aged SAC0307-0.05Pr solder at 150 °C for 840 h, (**e**) SEM image of aged SAC0307-0.05Pr solder at 150 °C for 840 h.

Table 1. EDS analyzed results of related IMCs formed in the aged SAC0307-0.05Pr solder.

Element Position	Sn		Ag		Cu		O		Pr	
Unit	wt.%	At.%	wt.%	At.%	wt.%	At.%	wt.%	At.%	wt.%	At.%
Point 1	35.68	22.35	/	/	/	/	10.57	49.23	53.75	28.42
Point 2	58.40	42.26	2.41	1.92	38.43	51.71	0.76	4.11	/	/
Point 3	26.80	23.8	71.98	70.43	0.45	0.75	0.77	5.02	/	/
Point 4	55.95	40.11	2.80	2.21	40.57	54.08	0.68	3.6	/	/
Point 5	30.68	27.69	68.59	68.21	0.15	0.26	0.58	3.84	/	/
Point 6	31.03	28.44	68.32	69.01	0.37	0.63	0.28	1.92	/	/

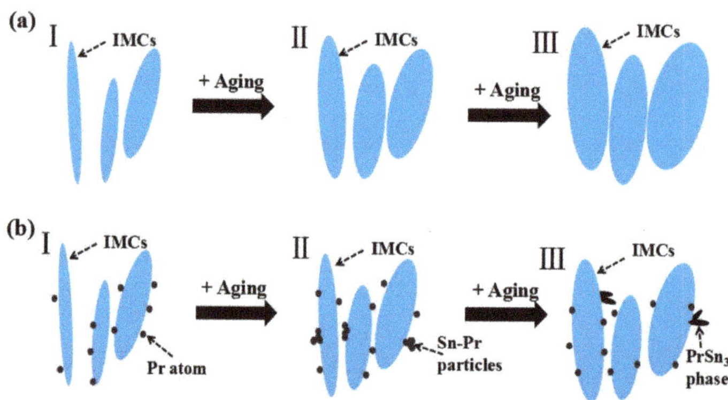

Figure 2. Illustration of changes in IMCs in (**a**) SAC0307 solder and (**b**) SAC0307-0.05Pr solder during aging process.

Figures 3 and 4 show changes of interfacial microstructures of SAC0307/Cu and SAC0307-0.05Pr/Cu joint subjected to different length of time of aging process. For the interfacial microstructure of SAC0307/Cu solder, a continuous hill-like Cu_6Sn_5 interfacial IMC layer [31] with sharp protrusions emerged after initial 72-h aging (Figure 3a). Moreover, a darker interfacial IMC layer, as identified to be Cu_3Sn [32] also grew near the Cu substrate. As aging time prolongs to 840 h, the radius of Cu_6Sn_5 increased with the number decreasing (Figure 3b–d). In addition, the thickness of Cu_3Sn also increased, and a detailed calculation is conducted in next section. In contrast to interfacial IMCs' growth at SAC0307/Cu interface with aging, Cu_6Sn_5 IMCs at SAC0307-0.05Pr/Cu interface were developed without any sharp Cu_6Sn_5 protrusions and the interfacial IMC layer was much more flat than the original aged one, even with aging treatment (Figure 4a–d). In addition, the growth of Cu_3Sn was also much slower than that at the SAC0307-0.05Pr/Cu interface. Figure 5 provides the changes in interfacial IMCs at both SAC0307 solder/Cu interface and SAC0307-0.05Pr solder/Cu interface during the aging process. Without Pr addition, the interfacial IMCs grows at a certain rate depending on the diffusions of Sn atoms from solder and Cu atoms from Cu substrate to the solder/Cu_6Sn_5 or Cu_6Sn_5/Cu_3Sn and Cu_3Sn/Cu interface. Hence, the shape of interfacial Cu_6Sn_5 IMCs varied with aging due to different diffusion rate of atoms (Figure 5(aI–III)). However, with a minute amount of Pr addition, the interfacial Cu_6Sn_5 IMCs became flatter due to the Pr atoms adsorbing on the surface of Cu_6Sn_5 IMCs, hindering their growth, as shown in Figure 5(bI). With aging going on, the diffusion of Sn and Cu atoms were largely affected by RE Pr atoms and the shape of interfacial IMC layer remains flat, as illustrated in Figure 5(bII,III).

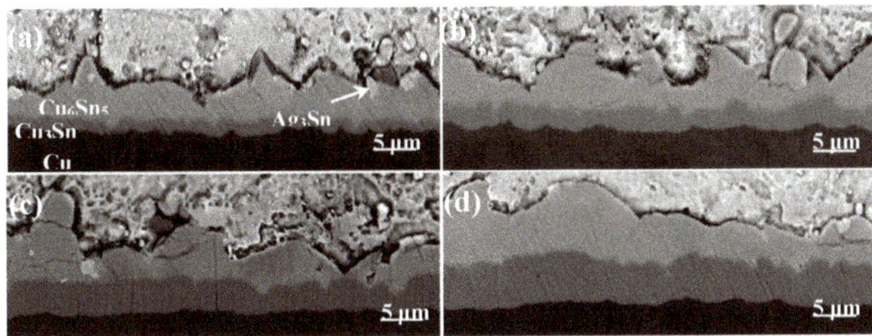

Figure 3. SEM images of the interfacial morphology at SAC0307 solder/Cu interfaces aged at 150 °C for different hrs: (**a**) 72 h; (**b**) 240 h; (**c**) 528 h; (**d**) 840 h.

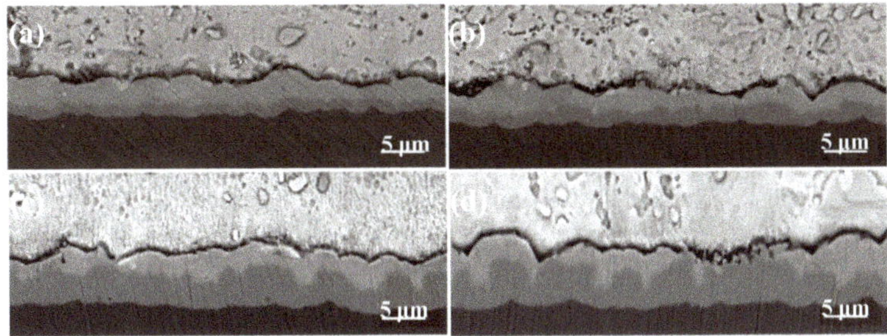

Figure 4. SEM images of the interfacial morphology at SAC0307-0.05Pr solder/Cu interfaces aged at 150 °C for different hrs: (**a**) 72 h; (**b**) 240 h; (**c**) 528 h; (**d**) 840 h.

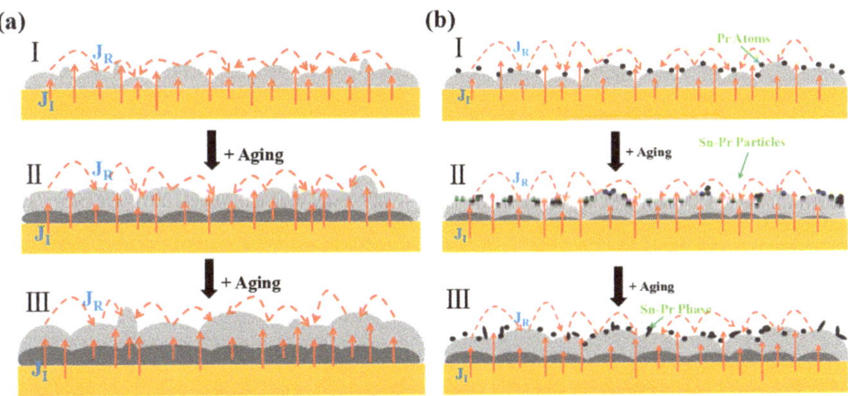

Figure 5. Illustration of changes in interfacial IMCs at (**a**) SAC0307 solder/Cu interface and (**b**) SAC0307-0.05Pr solder/Cu interface during aging process.

3.2. Thickness Changes of Interfacial IMC Layer during Isothermal Aging

For interfacial IMC layers at both interfaces of SAC0307/Cu and SAC0307-0.05Pr/Cu, the thickness of Cu_6Sn_5 and Cu_3Sn IMC layers raised with aging time, as summarized in Table 2. With 840 h aging, the total interfacial IMC layer thickness at SAC0307/Cu interface grew to 10.2 μm, whereas that at the SAC0307-0.05Pr/Cu increased only to 6.5 μm. In addition, Cu_3Sn IMC layers' thicknesses at these two types of interfaces reached 4.9 μm and 3.9 μm, respectively. Obviously, a minute amount of Pr effectively hindered the growth of interfacial IMCs. Commonly, the growth and development of interfacial IMC layer is diffusion-controlled. The relationship between interfacial IMC layer thickness and aging time meets the following classic diffusion formula [16]:

$$x_t = x_0 + \sqrt{Dt} \tag{1}$$

where x_t is interfacial IMC layer's thickness change after one period of aging time t, x_0 is its initial thickness, D is the growth coefficient, as calculated from the slope of the fitted curve of interfacial IMC layer thickness with aging time, as shown in Figure 6. The following lists the final results.

$$T_{(SAC0307\text{-}Total)} = 5.04 + 0.18\sqrt{t};\ T_{(SAC0307\text{-}Cu_6Sn_5)} = 4.28 + 0.03\sqrt{t};$$
$$T_{(SAC0307\text{-}Cu_3Sn)} = 0.76 + 0.14\sqrt{t}; \tag{2}$$

$$T_{(SAC0307\text{-}0.05Pr\text{-}Total)} = 2.67 + 0.13\sqrt{t};\ T_{(SAC0307\text{-}0.05Pr\text{-}Cu_6Sn_5)} = 2.31 + 0.005\sqrt{t};$$
$$T_{(SAC0307\text{-}0.05Pr\text{-}Cu_3Sn)} = 0.35 + 0.12\sqrt{t}; \tag{3}$$

Table 2. Detailed thickness change with aged time at the interfaces of solders/Cu.

Aging Hour (h)		72	240	528	840
Solder	**Interfacial IMC Layer**	Thickness (µm)			
SAC0307	Total (Cu_6Sn_5+Cu_3Sn)	6.42	8.04	8.9	10.2
	Cu_3Sn	1.93	3.05	3.96	4.9
	Cu_6Sn_5	4.49	4.99	4.94	5.3
SAC0307-0.05Pr	Total (Cu_6Sn_5+Cu_3Sn)	3.9	4.47	5.56	6.5
	Cu_3Sn	1.47	2.11	3.34	3.9
	Cu_6Sn_5	2.43	2.36	2.22	2.6

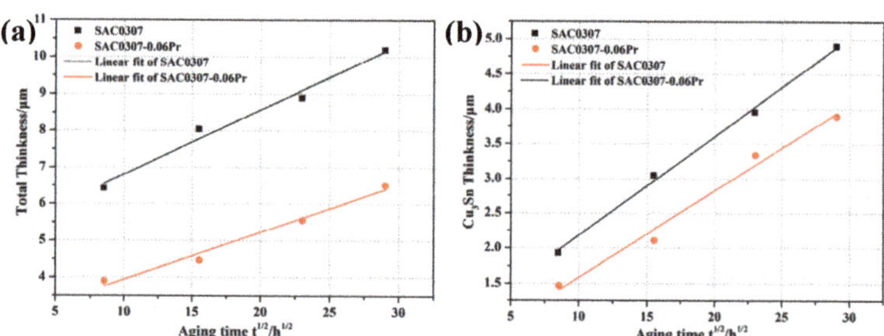

Figure 6. Thickness of the (**a**) total interfacial IMC layers (IMLs) and (**b**) Cu_3Sn IML at (**a**) SAC0307 solder/Cu interface and (**b**) SAC0307-0.05Pr solder/Cu interface aged at 150 °C for different hrs.

It can be calculated that the growth constant of total interfacial IMC layer (D_T), Cu_6Sn_5 (D_{Cu6}), Cu_3Sn (D_{Cu3}) for the SAC0307/Cu joint was 3.11×10^{-10} cm^2/s, 0.12×10^{-10} cm^2/s and 2.02×10^{-10} cm^2/s, respectively, while that for SAC0307-0.05Pr/Cu joint was 1.65×10^{-10} cm^2/s, 0.0021×10^{-10} cm^2/s (approximate to 0), and 1.53×10^{-10} cm^2/s, respectively. Obviously, doping 0.05 wt.% Pr can evidently lower the growth constant of Cu_6Sn_5 (D_{Cu6}), while the growth constant of Cu_3Sn (D_{Cu3}) was decreased slightly. The decreased growth constant of Cu_6Sn_5 (D_{Cu6}) was mainly due to Pr atoms pinning on the growth of Cu_6Sn_5. The suppressed mechanism for interfacial Cu_3Sn was mainly associated with a decreased gradient of Sn atom concentration at Cu_6Sn_5/Cu_3Sn interface. In addition, it can be observed that the decreased level for the growth constant of interfacial Cu_6Sn_5 (D_{Cu6}) is much larger than that of D_{Cu3}. This can be explained as follows. Generally, there are two ways to form interfacial Cu_6Sn_5. One is forming directly through Sn and Cu atom diffusion.

$$6Cu + 5Sn \rightarrow Cu_6Sn_5 \tag{4}$$

The other is through sacrificing interfacial Cu_3Sn,

$$2Cu_3Sn + 3Sn \rightarrow Cu_6Sn_5 \tag{5}$$

For producing interfacial Cu_3Sn IMCs, two approaches also can be reached. One is also forming directly:

$$3Cu + Sn \rightarrow Cu_3Sn \tag{6}$$

The other is developing through consuming interfacial Cu_6Sn_5 IMCs:

$$Cu_6Sn_5 + 9Cu \rightarrow 5Cu_3Sn \tag{7}$$

It can be seen that for directly forming interfacial Cu_6Sn_5 and Cu_3Sn IMCs (Equations (4) and (6)), per unit of Cu_6Sn_5 requires more Sn atoms than per unit of Cu_3Sn. For indirectly forming them (Equations (5) and (7)), Cu_3Sn just needs abundant supply of Cu atoms, whereas that of Cu_6Sn_5 still needs the supply of Sn atoms. Therefore, it can be concluded that what controls Cu_6Sn_5 IMCs' growth is the Sn atom diffusion and the solder/Cu_6Sn_5 interface state, while Cu atom diffusion determines the interface state of Cu_3Sn/Cu_6Sn_5 and determines the development of Cu_3Sn. Obviously, most Sn source comes from Sn-based solder, while most Cu source is from Cu substrate. Although most Pr atoms were adsorbed on the Cu substrate, hindering Cu diffusion in the initial reaction period, its inhibition effect was mainly on Sn atom diffusion with the growth of Cu_6Sn_5. Therefore, the interfacial Cu_6Sn_5 IMC growth was affected more than Cu_3Sn IMCs.

3.3. Shear Strength and Fracture Morphology after Aging Process

Figure 7a provides the corresponding illustration of shearing process and Figure 7b is the detailed plot between shear strength and aging time. Clearly, it can be observed that the shear strength of SAC0307-0.05Pr/Cu decreases with aging time at a certain rate, while the shear strength of SAC0307/Cu decreases also with a certain rate first but at a rapid decrease with aging time. After 840 h aging, the shear strengths of SAC0307/Cu and SAC0307-0.05Pr/Cu were declined to 9.5 MPa and 7.5 MPa, respectively. In addition, the shear strength of SAC0307-0.05Pr/Cu solder joint was larger than that of SAC0307/Cu solder joint after each aging process. This was mainly associated with their interfacial IMC layer' growth, as explained above.

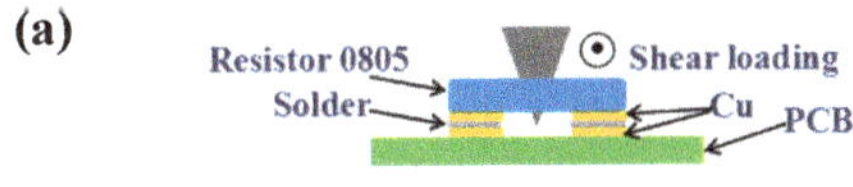

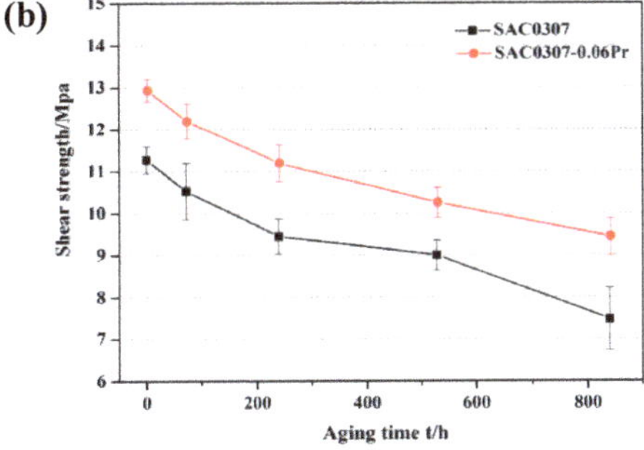

Figure 7. (**a**) Illustration of shearing process; (**b**) Changes in shear strength at Sn-0.3Ag-0.7Cu solder/Cu and Sn-0.3Ag-0.7Cu-0.05Pr/Cu solder aged for different hrs.

Figure 8 provides SEM images of corresponding fracture morphology at SAC0307/Cu, SAC0307-0.05Pr/Cu interfaces and the aged samples. Clearly, elongated dimples parallel to shear direction can be observed on the surface of fractured SAC0307/Cu shear joint, indicative of a distinct ductile fracture. For a fractograph of SAC0307-0.05Pr/Cu shear joint, dimples with smaller size emerge, indicating better ductility. After 840 h aging treatment, the dimples of these two types of joints became larger and more superficial, indicating the decrease in ductility with aging treatment. However, the average size of dimples on surface of fractured SAC0307-0.05Pr/Cu shear joint was still a bit larger than those on the fracture surface of SAC0307/Cu shear joint. It means the ductility of SAC0307-0.05Pr was still better than SAC0307/Cu solder even with 840 h aging. However, it should be noted that Pr oxides emerges on the fracture surface of SAC0307-0.05Pr/Cu shear joint after 840 h aging, possibly due to the appearance of PrSn₃ in the aged SAC0307-0.05Pr solder microstructure (Figure 1d).

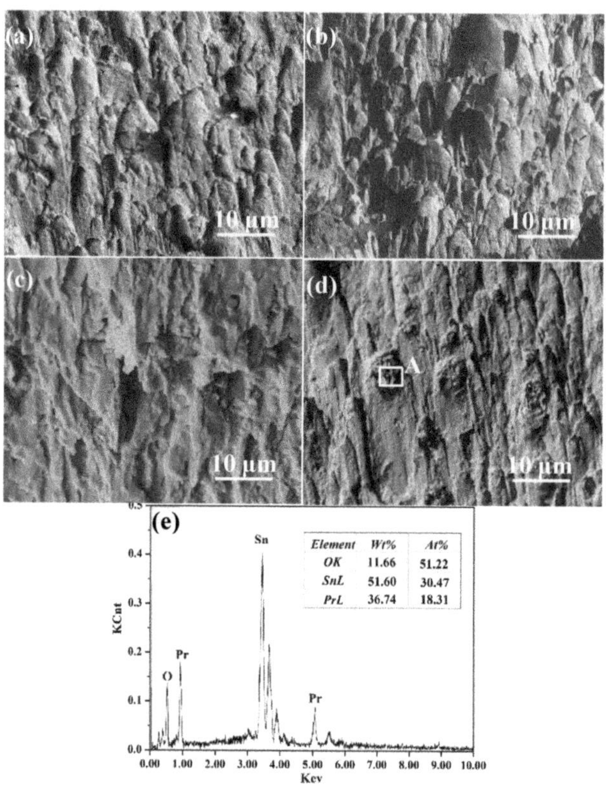

Figure 8. Fracture morphology on the surface of (**a**) SAC0307/Cu, (**b**) SAC0307-0.05Pr/Cu interface, and (**c**) aged SAC0307/Cu interface at 150 °C for 840 h, (**d**) aged SAC0307-0.05Pr/Cu interface for 840 h, (**e**) EDS data of area A in Figure 8d.

4. Conclusions

The effect of 0.05 wt.% Pr on high-temperature microstructure and shear strength of Sn-0.3Ag-0.7Cu/Cu low-Ag solder joint was studied and the following conclusions were obtained:

1. SAC0307-0.05Pr solder has a more refined microstructure with smaller size of Cu6Sn5 IMCs distributed on Sn matrix after 840 h aging when compared to Sn-0.3Ag-0.7Cu solder. This is because of the inhibiting effect on IMCs' growth by Pr atoms adsorbing on their grain surfaces.

2. During aging process, two types of interfacial IMC layers (Cu6Sn5 + Cu3Sn) were gradually developed at solder/Cu interfaces and the thicknesses of them also increased with aging time. Theoretical analysis indicated that doping 0.05 wt.% Pr evidently lowered the growth constant of Cu6Sn5(DCu6), while the growth constant of Cu3Sn (DCu3) decreased slightly.

3. SAC0307-0.05Pr/Cu solder joint has a higher shear strength than SAC0307/Cu solder joint even after 840 h aging. In this case, SAC0307-0.05Pr/Cu solder joint still has a better ductility than SAC0307/Cu.

Author Contributions: Investigation, J.W. (Jie Wu); Writing Original Draft Preparation, J.W. (Jie Wu); Writing, Review & Editing, S.X.; J.W. (Jingwen Wang); Supervision, S.X.; G.H.; Project Administration, J.W. (Jingwen Wang); Funding Acquisition, S.X.

Funding: This research was funded by [Songbai Xue] grant number [No. 51675269], by [Songbai Xue] grant number [No. 61605045] and by [Jie Wu] grant number [PAPD].

Acknowledgments: This project is supported by National Natural Science Foundation of China (Grant No. 51675269), National Natural Science Foundation of China (No. 61605045) and the Priority Academic Program Development of Jiangsu Higher Education Institutions (PAPD).

Conflicts of Interest: We declare no conflict of interest.

References

1. Zhang, L.; Han, J.-G.; Guo, Y.; He, C.-W. Anand model and FEM analysis of SnAgCuZn lead-free solder joints in wafer level chip scale packaging devices. *Microelectron. Reliab.* **2014**, *54*, 281–286. [CrossRef]

2. Tu, P.; Chan, Y.; Hung, K.; Lai, J.; Lai, J.K.L. Growth kinetics of intermetallic compounds in chip scale package solder joint. *Scr. Mater.* **2001**, *44*, 317–323. [CrossRef]

3. Cheng, B.; Wang, L.; Zhang, Q.; Gao, X.; Xia, X.; Kempe, W. Flip chip solder joint reliability under harsh environment. *Solder. Surf. Mt. Technol.* **2003**, *15*, 15–20.

4. Wu, J.; Xue, S.; Wang, J.-W.; Wang, J.-X.; Deng, Y. Enhancement on the high-temperature joint reliability and corrosion resistance of Sn-0.3Ag-0.7Cu low-Ag solder contributed by Al_2O_3 Nanoparticles (0.12 wt%). *J. Mater. Sci. Mater. Electron.* **2018**, *29*, 19663–19677.

5. Chen, W.-H.; Yu, C.-F.; Cheng, H.-C.; Tsai, Y.-M.; Lu, S.-T. IMC growth reaction and its effects on solder joint thermal cycling reliability of 3D chip stacking packaging. *Microelectron. Reliab.* **2013**, *53*, 30–40. [CrossRef]

6. Wu, J.; Xue, S.-B.; Wang, J.-W.; Liu, S.; Han, Y.-L.; Wang, L.-J. Recent progress of Sn–Ag–Cu lead-free solders bearing alloy elements and nanoparticles in electronic packaging. *J. Mater. Sci. Mater. Electron.* **2016**, *27*, 12729–12763. [CrossRef]

7. Huang, M.-L.; Zhao, N.; Liu, S.; He, Y.-Q. Drop failure modes of Sn–3.0Ag–0.5Cu solder joints in wafer level chip scale package. *Trans. Nonferrous Met. Soc. China* **2016**, *26*, 1663–1669. [CrossRef]

8. Mi, J.; Li, Y.-F.; Yang, Y.-J.; Peng, W.; Huang, H.-Z. Thermal Cycling Life Prediction of Sn-3.0Ag-0.5Cu Solder Joint Using Type-I Censored Data. *Sci. World J.* **2014**, *2014*. [CrossRef]

9. Zhang, L.; Tu, K. Structure and properties of lead-free solders bearing micro and nano particles. *Mater. Sci. Eng. R Rep.* **2014**, *82*, 1–32. [CrossRef]

10. Gain, A.K.; Zhang, L. Interfacial microstructure, wettability and material properties of nickel (Ni) nanoparticle doped tin-bismuth-silver (Sn-Bi-Ag) solder on copper (Cu) substrate. *J. Mater. Sci. Mater. Electron.* **2016**, *27*, 3982–3994. [CrossRef]

11. Gnecco, F.; Ricci, E.; Amore, S.; Giuranno, D.; Borzone, G.; Zanicchi, G.; Novakovic, R. Wetting behaviour and reactivity of lead free Au-In-Sn and Bi-In-Sn alloys on copper substrates. *Int. J. Adhes. Adhes.* **2007**, *27*, 409–416. [CrossRef]

12. Gain, A.K.; Zhang, L. Harsh service environment effects on the microstructure and mechanical properties of Sn–Ag–Cu-1 wt% nano-Al solder alloy. *J. Mater. Sci. Mater. Electron.* **2016**, *27*, 11273–11283. [CrossRef]

13. Wang, Y.; Lin, Y.; Tu, C.; Kao, C. Effects of minor Fe, Co, and Ni additions on the reaction between SnAgCu solder and Cu. *J. Alloys Compd.* **2009**, *478*, 121–127. [CrossRef]

14. Gain, A.K.; Zhang, L. Microstructure, mechanical and electrical performances of zirconia nanoparticles-doped tin-silver-copper solder alloys. *J. Mater. Sci. Mater. Electron.* **2016**, *27*, 7524–7533. [CrossRef]

15. Gain, A.K.; Zhang, L.; Chan, Y.C. Microstructure, elastic modulus and shear strength of alumina (Al_2O_3) nanoparticles-doped tin-silver-copper (Sn-Ag-Cu) solders on copper (Cu) and gold/nickel (Au/Ni)-plated Cu substrates. *J. Mater. Sci. Mater. Electron.* **2015**, *26*, 7039–7048. [CrossRef]

16. Tsao, L.C.; Chang, S.Y. Effects of Nano-TiO$_2$ additions on thermal analysis, microstructure and tensile properties of Sn$_{3.5}$Ag$_{0.25}$Cu solder. *J. Mater. Sci. Mater Electron.* **2015**, *26*, 7039–7048. [CrossRef]

17. Li, B.; Shi, Y.; Lei, Y.; Guo, F.; Xia, Z.; Zong, B. Effect of rare earth element addition on the microstructure of Sn-Ag-Cu solder joint. *J. Electron. Mater.* **2015**, *34*, 217–224. [CrossRef]

18. Xiao, W.; Shi, Y.; Xu, G.; Ren, R.; Guo, F.; Xia, Z.; Lei, Y. Effect of rare earth on mechanical creep–fatigue property of SnAgCu solder joint. *J. Alloys Compd.* **2009**, *472*, 198–202. [CrossRef]

19. Wu, J.; Xue, S.; Wang, J.; Wang, J.; Liu, S. Effect of Pr addition on properties and Sn whisker growth of Sn–0.3Ag–0.7Cu low-Ag solder for electronic packaging. *J. Mater. Sci. Mater. Electron.* **2017**, *37*, 1640–10244. [CrossRef]

20. Sadiq, M.; Pesci, R.; Cherkaoui, M. Impact of Thermal Aging on the Microstructure Evolution and Mechanical Properties of Lanthanum-Doped Tin-Silver-Copper Lead-Free Solders. *J. Electron. Mater.* **2013**, *42*, 492–501. [CrossRef]

21. Haseeb, A.; Leng, T.S. Effects of Co nanoparticle addition to Sn–3.8Ag–0.7Cu solder on interfacial structure after reflow and ageing. *Intermetallics* **2011**, *19*, 707–712. [CrossRef]

22. Gu, Y.; Zhao, X.; Li, Y.; Liu, Y.; Wang, Y.; Li, Z. Effect of nano-Fe$_2$O$_3$ additions on wettability and interfacial intermetallic growth of low-Ag content Sn–Ag–Cu solders on Cu substrates. *J. Alloys Compd.* **2015**, *627*, 39–47. [CrossRef]

23. Wu, J.; Xue, S.; Wang, J.; Wu, M.; Wang, J. Effects of α-Al$_2$O$_3$ nanoparticles-doped on microstructure and properties of Sn–0.3Ag–0.7Cu low-Ag solder. *J. Mater. Sci. Mater. Electron.* **2018**, *29*, 7372–7387. [CrossRef]

24. Wu, J.; Xue, S.; Wang, J.; Wu, M. Coupling effects of rare-earth Pr and Al$_2$O$_3$ nanoparticles on the microstructure and properties of Sn-0.3Ag-0.7Cu low-Ag solder. *J. Alloys Compd.* **2019**, *784*, 471–487. [CrossRef]

25. Yu, J.; Welford, R.; Hills, P. Industry responses to EU WEEE and ROHS directives: Perspectives from China. *Corp. Soc. Responsib. Environ. Manag.* **2012**, *13*, 286–299. [CrossRef]

26. Kotadia, H.R.; Howes, P.D.; Mannan, S.H. A review: On the development of low melting temperature Pb-free solders. *Microelectron. Reliab.* **2014**, *54*, 1253–1273. [CrossRef]

27. Gain, A.K.; Zhang, L. High-temperature and humidity change the microstructure and degrade the material properties of tin-silver interconnect material. *Microelectron. Reliab.* **2018**, *83*, 101–110. [CrossRef]

28. Cheng, S.; Huang, C.-M.; Pecht, M. A review of lead-free solders for electronics applications. *Microelectron. Reliab.* **2017**, *75*, 77–95. [CrossRef]

29. Gain, A.K.; Zhang, L. Growth nature of in-situ Cu$_6$Sn$_5$-phase and their influence on creep anddamping characteristics of Sn-Cu material under high-temperature and humidity. *Microelectron. Reliab.* **2018**, *87*, 278–285. [CrossRef]

30. Chuang, T.; Lin, H. Size Effect of Rare-Earth Intermetallics in Sn-9Zn-0.5Ce and Sn-3Ag-0.5Cu-0.5Ce Solders on the Growth of Tin Whiskers. *Met. Mater. Trans. A* **2008**, *39*, 2862–2866. [CrossRef]

31. Huang, J.Q.; Zhou, M.B.; Liang, S.B.; Zhang, X.P. Size effects on the interfacial reaction and microstructural evolution of Sn-ball/Sn3.0Ag0.5Cu-paste/Cu joints in board-level hybrid BGA interconnection at critical reflowing temperature. *J. Mater. Sci. Mater. Electron.* **2018**, *29*, 7651–7660. [CrossRef]

32. Yao, P.; Li, X.; Liang, X.; Yu, B. Investigation of soldering process and interfacial microstructure evolution for the formation of full Cu$_3$Sn joints in electronic packaging. *Mater. Sci. Semicond. Process.* **2017**, *58*, 39–50. [CrossRef]

 *applied sciences*

Article

Enhanced Tensile Plasticity in Ultrafine Lamellar Eutectic Al-CuBased Composites with α-Al Dendrites Prepared by Progressive Solidification

Jialin Cheng [1,2,*], Yeling Yun [1] and Jiaxin Rui [1]

[1] School of Materials Science and Engineering, Nanjing Institute of Technology, Nanjing 211167, China; yunyeling1994@163.com (Y.Y.); NJIT_Rjx@163.com (J.R.)

[2] Jiangsu Key Laboratory of Advanced Structural Materials and Application Technology, Nanjing 211167, China

* Correspondence: chengjialin@njit.edu.cn

Received: 9 August 2019; Accepted: 15 September 2019; Published: 19 September 2019

Featured Application: The developed Al-Cu ultrafine lamellar eutectic composites with excellent mechanical properties have potential widely applications in the realms of aviation, aerospace and automotive.

Abstract: In this paper, a new class of Al-Cubased composites which combine the ultrafine lamellar eutectic matrix (α-Al + θ-Al$_2$Cu) and micron-sized primary α-Al dendrites was prepared by progressive solidification. By adjusting the alloy composition and solidification process, the formation of favorable microstructural and micromechanical features can be achieved. The ultrafine lamellar eutectic composite Al$_{94}$Cu$_6$ exhibits excellent mechanical properties with 472 MPa fracture strength and 7.4% tensile plastic strain. The plasticity of the ultrafine lamellar eutectic composite relies on the volume fraction and work hardening ability of micron-scale primary phase. The present results provide a new perspective for improving the plasticity of the ultrafine lamellar eutectic alloys.

Keywords: ultrafine lamellar eutectic structure; composite; plasticity

---

1. Introduction

Bulk nanocrystalline alloys have been highlighted since the first report by Gleiter et al. [1] in the 1980s, because of their high strength and low elastic modulus in comparison with conventional coarse-polycrystalline alloys. Several synthesis methods for the bulk nanocrystalline alloys have been developed, such as powder consolidation [2], amorphous crystallization [3], severe plastic deformation [4] and electrodeposition [5–7]. However, these methods have multiple processing steps and are not easily commercially viable. Recently, some nano/ultrafine lamellar eutectic alloys have been developed just using the simple and low-cost single step casting [8–10] and have attracted much attention for both significant science and engineering interests. However, like the most bulk nano-structured alloys and bulk metallic glasses (BMGs), these nano/ultrafine lamellar eutectic alloys usually fail catastrophically at ambient temperature by the highly localized deformation behavior, which severely restricts their commercial application as structural materials.

To block the high localized shear deformation, some inhomogeneous microstructures such as the micron-scale soft and ductile dendritic phase have been introduced into the nano/ultrafine matrix [11–19]. Although these nano/ultrafine structured composites exhibited an excellent compressive plasticity, they still exhibited very limited macroscopic plasticity under tensile stress. For example [19], the compressive plasticity of the Ti-based alloys with ultrafine lamellar eutectic structure is as high as 30%, while the tensile plasticity is less than 1%. Similar to the nano/ultrafine structured composites, the BMG composites containing in situ soft and ductile dendrites were developed in 2000, which also exhibit high compressive plasticity and very limit tensile plasticity [20]. In 2008, Hofmann et al. [21] made a breakthrough in tensile plasticity in BMG composites. They proposed two basic principles based on matching fundamental micromechanical characteristics and microstructural length scales to design the ductile BMG composites, which were: (1) introduction of 'soft' elastic/plastic inhomogeneities to initiate local shear banding; and (2) matching of microstructural length scales to the characteristic length scale R_P (plastic shielding of an opening crack tip) to suppress the instability propagation of shear bands and micro-cracks. Subsequently, numerous ductile BMG composites with large tensile plasticity were developed [22–26].

The authors proposed that these principles are applicable to the nanocrystalline and ultrafine lamellar eutectic alloys. In this study, we select the simple Al-Cu binary alloy system, which has significant scientific and commercial interests due to its high specific strength and relatively low cost. By matching the characteristics of the in situ ductile dendrites, including size, volume fraction and hardness, the bulk Al-Cu ultrafine lamellar eutectic composites with enhanced tensile plasticity were developed using simple casting. The effects of the microstructure and micromechanical features on the macromechanical properties of ultrafine eutectic composites are also discussed in detail. Our current findings give a new clue for developing nanoscale or ultrafine-grained composites with excellent mechanical properties.

2. Experimental Procedures

The master alloys of $Al_{83}Cu_{17}$, $Al_{90}Cu_{10}$ and $Al_{94}Cu_6$ (at. %) were prepared from the Al and Cu pieces with industrial purities of 99.2 (wt. %) by arc-melting. The master alloys were then cast into Cu molds to form the rod-shaped samples with 7 mm diameter. To minimize the cast flaws, these rods were remelted and progressively solidified at a withdrawal velocity of 4.0 mm/s under the directional solidification device. The temperature gradient was about 17 K/mm.

X-ray diffractometry (XRD) and optical microscopy (OM) were used to observe the microstructure. The element distribution in the solidified microstructures was determined by energy dispersive X-ray spectrometer (EDS) attached to the scanning electron microscopy (SEM) JSM-6380LV (JEOL Ltd., Akishima, Tokyo and Japan). The tensile specimens with a 12.4 mm gauge length and 2.5 mm diameter were machined and tested on an Instron-8801 testing machine under quasistatic loading at an initial engineering strain rate of $5 \times 10^{-4} \cdot s^{-1}$. A CSM-NHT2 (CSM Instruments, Peuseux and Switzerland) nano-indentation instrument was applied to investigate the hardness and elastic modulus (E) of composites. The nano-indentation tests were loaded to 50 mN and kept for 10 s. Each sample was measured at least five times to ensure that the results are reproducible and statistically meaningful. The fracture surfaces were carefully observed though SEM.

3. Results and Discussion

The XRD patterns and optical microscopies of the alloys are presented in Figure 1. As shown in Figure 1a, the three alloys show very similar diffraction patterns, indicating that all the alloys are composed of α-Al solid solution and θ-Al_2Cu phase. However, the microstructures of alloys are obviously different. The $Al_{83}Cu_{17}$ alloy exhibits a typical ultrafine lamellar eutectic microstructure, in which the white α-Al and black θ-Al_2Cu are arranged alternately. The lamellar spacing is about 0.6 µm, as show in Figure 1b. In a previous study, Park et al. [27] reported that the lamellar spacing was 0.2–0.3 µm for the $Al_{83}Cu_{17}$ alloy which was only for 1 mm diameter rod-shaped samples prepared by Cu mold casting. Obviously, the cooling rate in present study is much lower than previously reported. However, the increase in lamellar spacing is not very significant, only from 0.2–0.3 µm to 0.6 µm, which is very beneficial to industrial production.

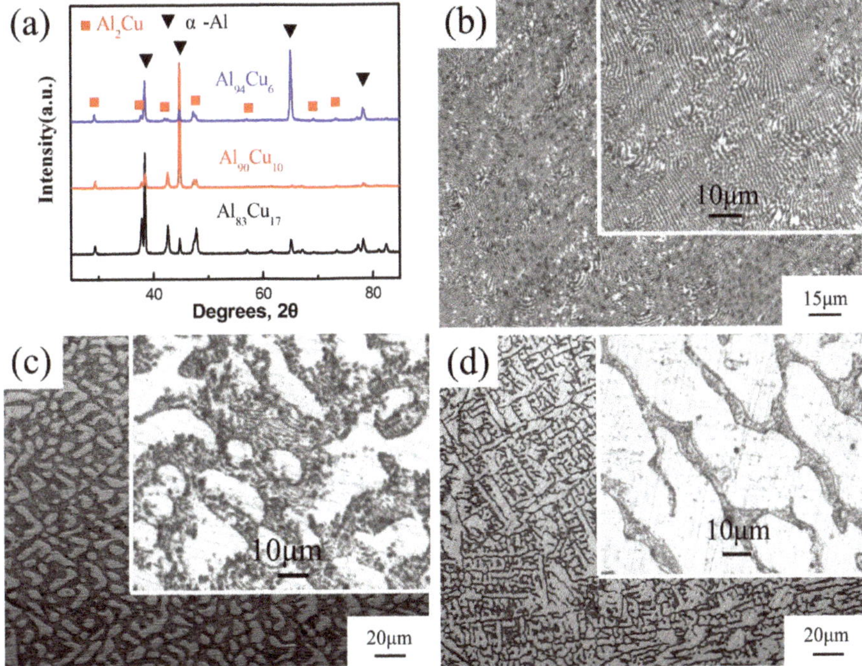

Figure 1. (**a**) X-ray diffractometry (XRD) patterns of alloys and their optical microscopy (OM) micrographs of the (**b**) $Al_{83}Cu_{17}$, (**c**) $Al_{90}Cu_{10}$ and (**d**) $Al_{94}Cu_6$. The up-insets in (**b**–**d**) show their magnified views of the corresponding microstructures.

The hypoeutectic $Al_{90}Cu_{10}$ alloy shows a typical composite structure, in which white particles are uniformly embedded in the eutectic matrix. According to the EDS analysis (see Figure 2), the primary phase is enriched in Al and can be identified as the α-Al phase with the average compositions of $Al_{92.9}Cu_{7.1}$. The volume fraction and average grain size of the α-Al phase in the $Al_{90}Cu_{10}$ alloy is about 44% and 7 µm, respectively. It is noteworthy that the eutectic matrix is still composed of the ultrafine lamellar α-Al + θ-Al_2Cu eutectic microstructure, and the lamellar spacing is similar to the $Al_{83}Cu_{17}$ alloy. For the $Al_{94}Cu_6$ alloy, the volume fraction of α-Al phase increases to 74%, and the grain morphology changes from particle to the dendrite. While the average compositions of the primary α-Al phase is $Al_{97.8}Cu_{2.2}$ in the $Al_{94}Cu_6$ alloy, which obviously has a lower Cu content than that of the $Al_{90}Cu_{10}$ alloy.

Figure 3 presents the engineering stress-strain curves and nano-indentation load-displacement curves of the alloys. The corresponding mechanical properties are summarized in Tables 1 and 2. As shown, the $Al_{83}Cu_{17}$ alloy exhibits the highest hardness and tensile strength, which are 2.5 GPa and 758 MPa, respectively, because of its completely ultrafine lamellar eutectic structure. However, the $Al_{83}Cu_{17}$ alloy fails catastrophically without any plasticity. In the previous study [27], the $Al_{83}Cu_{17}$ alloy with finer lamellar spacing also fails in a brittle manner under the room compressive tests. This brittle fracture is mainly due to the lack of a work hardening mechanism, resulting in highly localized deformation.

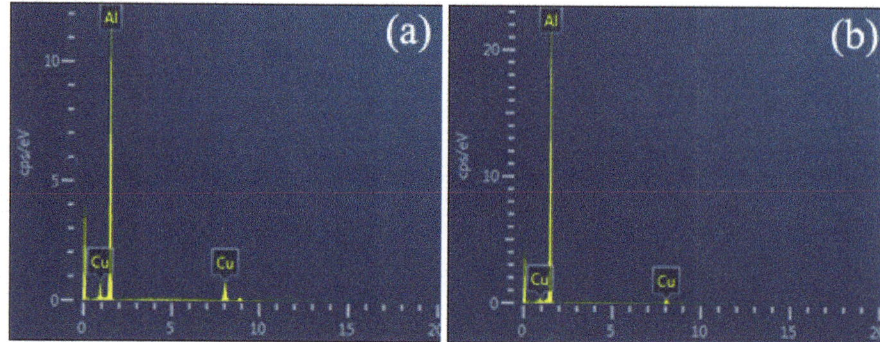

Figure 2. The SEM-EDS analysis of α-Al phase in alloys Al$_{90}$Cu$_{10}$ (**a**) and Al$_{94}$Cu$_6$ (**b**).

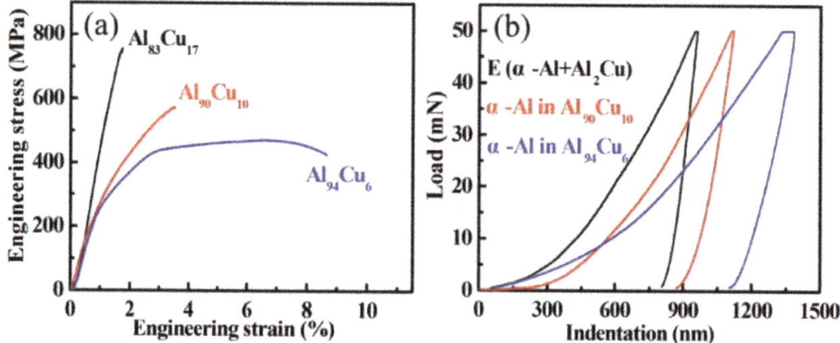

Figure 3. (**a**) Engineering tensile stress-strain curves of alloys, and (**b**) nano-indentation load-displacement curves of the ultrafine lamellar eutectic matrices and α-Al phase in alloys.

Table 1. Summary of yield strength (σ_y), tensile strength (σ_b) and tensile plastic strain (ε_p) of Al-based alloys.

Composition	σ_y (MPa)	σ_b (MPa)	ε_p (%)
Al$_{83}$Cu$_{17}$	-	758	-
Al$_{90}$Cu$_{10}$	392	571	2.0
Al$_{94}$Cu$_6$	306	472	7.4

Table 2. Hardness (H) and elastic modulus (E) of different structures in alloys, measured by nano-indentation.

Structure	H (GPa)	E (GPa)
Eutectic (α-Al + θ-Al$_2$Cu)	2.5 ± 0.2	86 ± 2
α-Al in Al$_{90}$Cu$_{10}$ alloy	1.8 ± 0.2	64 ± 3
α-Al in Al$_{94}$Cu$_6$ alloy	1.3 ± 0.2	40 ± 1

When some primary α-Al particles are precipitated on the ultrafine lamellar eutectic matrix as in the case of Al$_{90}$Cu$_{10}$ and Al$_{94}$Cu$_6$, the alloys present lower tensile strengths (Table 1). In general, like most composites, we proposed that the deformation behaviors of these ultrafine lamellar eutectic composites should follow a rule-of-mixtures relationship. According to the nano-indentation analysis, the hardness and E of the ultrafine eutectic matrix in the Al$_{90}$Cu$_{10}$ and Al$_{94}$Cu$_6$ alloys are almost the same as that of the $Al_{83}Cu_{17}$ alloy, due to the similar compositions and lamellar spacing of ultrafine eutectic matrix in the three alloys. While the hardness and E of the α-Al phase in the $Al_{90}Cu_{10}$ and $Al_{94}Cu_6$ alloys are much lower than that of the ultrafine eutectic matrix (Table 2). Therefore, the yield strength of the ultrafine lamellar eutectic composites $Al_{90}Cu_{10}$ and $Al_{94}Cu_6$ obviously decreases with the precipitation of the α-Al phase.

As expected, the precipitated α-Al phase can enhance the ductility of alloys. The $Al_{90}Cu_{10}$ alloy exhibits obvious work hardening and 2% tensile plastic strain, while the plastic strain of the $Al_{94}Cu_6$ alloy is significantly improved to 7.4%. Obviously, the volume fraction of the primary α-Al phase has significant effects on the plasticity of the composites. For high enough volume fractions, the precipitated α-Al dendrites not only bear more deformation in themselves, but also connect with each other and form a continuous network distribution of "hand-in-hand", which are beneficial to suppress the local instability propagation of shear bands and micro-cracks. In previous studies, Lee et al. [28] also found that there was a critical content of 40 vol% primary soft dendrites, beyond which the ductility of the La-based BMG composites escalates rapidly. Moreover, as show in Figure 3b and Table 2, the hardness and E of the α-Al phase in the $Al_{94}Cu_6$ alloy are much lower than that of the $Al_{90}Cu_{10}$ alloy, indicating it has much higher ductility and stronger work hardening ability. Under the process of deformation, the primary α-Al phase with higher work hardening ability can promote the redistribution of stress and avoid excessive stress concentration, thus delaying the plastic instability and obtaining larger plasticity. Xia et al. [29] also found that with the same α-Al volume fraction and size, the tensile plasticity of the Cu–Al alloys with bimodal structures was significantly enhanced by increasing the work hardening ability of the micron-scale primary phase. These results indicate that the microstructure (volume fraction, size and morphology) and micromechanical properties (hardness, E and work hardening ability) of the micron-scale primary phase have a significant effect on the plasticity of the ultrafine lamellar eutectic composite. More quantitative analysis and discussion on the relationships between the microstructure, micromechanical features and macroscopic properties should be further studied in the future.

Figure 4 shows the fracture surface morphologies. As shown, the macroscopic fracture surface of the $Al_{83}Cu_{17}$ alloy (inset in Figure 4a) shows cleavage-like features, and only a few main cracks penetrate the whole fracture surface, indicating a brittle fracture. The detailed microscopic observation evidently shows that the deformation of α-Al layers is accompanied with pull-out and softening, as indicated by white arrows in Figure 4a. While the θ-Al_2Cu layers fail by a predominantly faceted cleavage fracture. As shown in Figure 4b, ultrafine lamellar eutectic composite $Al_{90}Cu_{10}$ shows a complex fracture surface with many isolated dimple-like patterns and some cleavage-like features, which are fractures characteristic of the ductile α-Al and brittle ultrafine lamellar eutectic matrix, respectively. For the $Al_{94}Cu_6$ alloy, the main fracture feature is similar to the $Al_{90}Cu_{10}$ alloy, but much more dimple-like patterns connect with each other and form a continuous distribution. Moreover, an apparent necking can be observed, indicating more plastic deformation.

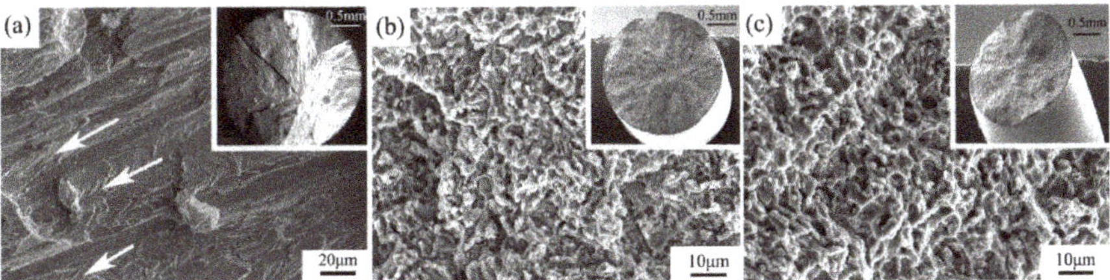

Figure 4. The microscopic fracture morphologies for alloys (**a**) $Al_{83}Cu_{17}$, (**b**) $Al_{90}Cu_{10}$ and (**c**) $Al_{94}Cu_6$ are shown. The insets in (**a–c**) show the macroscopic fracture surfaces of the corresponding alloys.

4. Conclusions

In conclusion, the ultrafine lamellar eutectic Al-Cu alloy and the composites with primary soft α-Al dendrites have been prepared by progressive solidification. Compared with the ultrafine lamellar eutectic alloy, the composites exhibit excellent mechanical properties combining high strength and large tensile plasticity. Moreover, the present results indicate that the microstructural and micromechanical features including volume fraction, distribution and hardness and E of the primary phase are very crucial to the mechanical properties of the ultrafine lamellar eutectic composites. These findings give a new clue for developing nanoscale or ultrafine-grained composites with excellent mechanical properties, especially in the binary or ternary ultrafine lamellar eutectic systems.

Author Contributions: J.C., Y.Y. and J.R.: Co-organized the work, prepared the materials, characterized the materials with OM, XRD and SEM, and wrote the manuscript draft.

Acknowledgments: This work was supported by Excellent Youth Foundation of Jiangsu Scientific Committee (BK20180106), the Qinglan Project of Jiangsu Province of China, the Innovation Funds of Nanjing Institute of Technology (Grant No. CKJA201703), the Innovative Foundation Project for Students of Jiangsu Province (201811276024), the Opening Project of Jiangsu Key Laboratory of Advanced Structural Materials and Application Technology (ASMA201705).

Conflicts of Interest: The authors declare no conflict of interest.

References

1. Gleiter, H. Nanocrystalline materials. *Prog. Mater. Sci.* **1989**, *33*, 223–315. [CrossRef]
2. Koch, C. The synthesis and structure of nanocrystalline materials produced by mechanical attrition: A review. *Nanostruct. Mater.* **1993**, *2*, 109–129. [CrossRef]
3. Lu, K. Nanocrystalline metals crystallized from amorphous solids: Nanocrystallization, structure, and properties. *Mater. Sci. Eng. R Rep.* **1996**, *16*, 161–221. [CrossRef]
4. Valiev, R.; Islamgaliev, R.; Alexandrov, I. Bulk nanostructured materials from severe plastic deformation. *Prog. Mater. Sci.* **2000**, *45*, 103–189. [CrossRef]
5. Erb, U.; El-Sherik, A.; Palumbo, G.; Aust, K. Synthesis, structure and properties of electroplated nanocrystalline materials. *Nanostruct. Mater.* **1993**, *2*, 383–390. [CrossRef]
6. Hughes, G.; Smith, S.; Pande, C.; Johnson, H.; Armstrong, R. Hall-petch strengthening for the microhardness of twelve nanometer grain diameter electrodeposited nickel. *Scr. Met.* **1986**, *20*, 93–97. [CrossRef]
7. Lu, L.; Sui, M.L. Superplastic Extensibility of Nanocrystalline Copper at Room Temperature. *Science* **2000**, *287*, 1463–1466. [CrossRef]
8. Nagarajan, R.; Chattopadhyay, K. Intermetallic Ti2Ni/TiNi nanocomposite by rapid solidification. *Acta Met. Mater.* **1994**, *42*, 947–958. [CrossRef]
9. He, G.; Eckert, J.; Löser, W.; Schultz, L. Novel Ti-base nanostructure-dendrite composite with enhanced plasticity. *Nat. Mater.* **2003**, *2*, 33–37. [CrossRef]
10. Louzguine, D.V.; Kato, H.; Inoue, A. High strength and ductile binary Ti–Fe composite alloy. *J. Alloys Compd.* **2004**, *384*, L1–L3. [CrossRef]
11. Park, J.; Kim, K.; Kim, D.; Mattern, N.; Li, R.; Liu, G.; Eckert, J. Multi-phase Al-based ultrafine composite with multi-scale microstructure. *Intermet.* **2010**, *18*, 1829–1833. [CrossRef]
12. Cao, G.; Peng, Y.; Liu, N.; Li, X.; Lei, Z.; Ren, Z.; Gerthsen, D.; Russell, A. Formation of a bimodal structure in ultrafine Ti–Fe–Nb alloys with high-strength and enhanced ductility. *Mater. Sci. Eng. A* **2014**, *609*, 60–64. [CrossRef]
13. Maity, T.; Das, J. High strength Ni–Zr–(Al) nanoeutectic composites with large plasticity. *Intermet.* **2015**, *63*, 51–58. [CrossRef]
14. Kim, J.; Hong, S.; Park, H.; Park, G.; Suh, J.-Y.; Park, J.; Kim, K. Influence of microstructural evolution on mechanical behavior of Fe–Nb–B ultrafine composites with a correlation to elastic modulus and hardness. *J. Alloys Compd.* **2015**, *647*, 886–891. [CrossRef]
15. Lei, Q.; Ramakrishnan, B.P.; Wang, S.; Wang, Y.; Mazumder, J.; Misra, A. Structural refinement and nanomechanical response of laser remelted Al-$Al_{2}Cu$ lamellar eutectic. *Mater. Sci. Eng. A* **2017**, *706*, 115–125. [CrossRef]

16. Ramakrishnan, B.P.; Lei, Q.; Misra, A.; Mazumder, J. Effect of laser surface remelting on the microstructure and properties of Al-Al2Cu-Si ternary eutectic alloy. *Sci. Rep.* **2017**, *7*, 13468. [CrossRef] [PubMed]

17. Kim, J.; Hong, S.; Park, J.; Eckert, J.; Kim, K. Microstructure and mechanical properties of hierarchical multi-phase composites based on Al-Ni-type intermetallic compounds in the Al-Ni-Cu-Si alloy system. *J. Alloys Compd.* **2018**, *749*, 205–210. [CrossRef]

18. Kim, J.; Hong, S.; Kim, Y.; Park, H.; Maity, T.; Chawake, N.; Bian, X.; Sarac, B.; Park, J.; Prashanth, K.; et al. Cooperative deformation behavior between the shear band and boundary sliding of an Al-based nanostructure-dendrite composite. *Mater. Sci. Eng. A* **2018**, *735*, 81–88. [CrossRef]

19. Yang, C.; Kang, L.; Li, X.; Zhang, W.; Zhang, D.; Fu, Z.; Li, Y.; Zhang, L.; Lavernia, E. Bimodal titanium alloys with ultrafine lamellar eutectic structure fabricated by semi-solid sintering. *Acta Mater.* **2017**, *132*, 491–502. [CrossRef]

20. Hays, C.C.; Kim, C.P.; Johnson, W.L. Microstructure Controlled Shear Band Pattern Formation and Enhanced Plasticity of Bulk Metallic Glasses Containing in situ Formed Ductile Phase Dendrite Dispersions. *Phys. Rev. Lett.* **2000**, *84*, 2901–2904. [CrossRef]

21. Hofmann, D.C.; Suh, J.Y.; Wiest, A.; Duan, G.; Lind, M.L.; Demetriou, M.D.; Johnson, W.L. Designing metallic glass matrix composites with high toughness and tensile ductility. *Nature* **2008**, *451*, 1085–1089. [CrossRef] [PubMed]

22. Liu, D.M.; Lin, S.F.; Ge, S.F.; Zhu, Z.W.; Fu, H.M.; Zhang, H.F. A Ti-based bulk metallic glass composite with excellent tensile properties and significant work-hardening capacity. *Mater. Lett.* **2018**, *233*, 107–110. [CrossRef]

23. Cheng, J.; Chen, G.; Zhao, W.; Wang, Z.; Zhang, Z. Enhancement of tensile properties by the solid solution strengthening of nitrogen in Zr-based metallic glass composites. *Mater. Sci. Eng. A* **2017**, *696*, 461–465. [CrossRef]

24. Cheng, J.; Wang, J.; Yun, Y.; Rui, J.; Zhao, W.; Li, F. A novel core-shell structure reinforced Zr-based metallic glass composite with combined high strength and good tensile ductility. *J. Alloys Compd.* **2019**, *803*, 413–416. [CrossRef]

25. Wu, Y.; Xiao, Y.; Chen, G.; Liu, C.T.; Lu, Z. Bulk Metallic Glass Composites with Transformation-Mediated Work-Hardening and Ductility. *Adv. Mater.* **2010**, *22*, 2770–2773. [CrossRef]

26. Song, W.; Wu, Y.; Wang, H.; Liu, X.; Chen, H.; Guo, Z.; Lu, Z. Microstructural Control via Copious Nucleation Manipulated by In Situ Formed Nucleants: Large-Sized and Ductile Metallic Glass Composites. *Adv. Mater.* **2016**, *28*, 8156–8161. [CrossRef] [PubMed]

27. Park, J.M.; Mattern, N.; Kühn, U.; Eckert, J.; Kim, K.B.; Kim, W.T.; Chattopadhyay, K.; Kim, D.H. High-strength bulk Al-based bimodal ultrafine eutectic composite with enhanced plasticity. *J. Mater. Res.* **2009**, *24*, 2605–2609. [CrossRef]

28. Lee, M.L.; Li, Y.; Schuh, C.A. Effect of a controlled volume fraction of dendritic phases on tensile and compressive ductility in La-based metallic glassmatrix composites. *Acta Mater.* **2004**, *52*, 4121–4131. [CrossRef]

29. Xia, S.H.; Wang, J.T. A micromechanical model of toughening behavior in the dual-phase composite. *Int. J. Plast.* **2010**, *26*, 1442–1460. [CrossRef]

 *applied sciences*

Article

Enhancing Properties of Aerospace Alloy Elektron 21 Using Boron Carbide Nanoparticles as Reinforcement

Sravya Tekumalla [1], Ng Joo Yuan [1], Meysam Haghshenas [2] and Manoj Gupta [1,*]

[1] Department of Mechanical Engineering, National University of Singapore, Singapore 117575, Singapore; tvrlsravya@u.nus.edu (S.T.); e0004161@u.nus.edu (N.J.Y.)

[2] Department of Mechanical, Industrial and Manufacturing Engineering, University of Toledo, Toledo, OH 43606, USA; Meysam.Haghshenas@UToledo.Edu

* Correspondence: mpegm@nus.edu.sg

Received: 12 September 2019; Accepted: 9 December 2019; Published: 12 December 2019

Featured Application: Aerospace sector.

Abstract: In this study, the effect of nano-B_4C addition on the property profile of Elektron 21 (E21) alloys is investigated. E21 reinforced with different amounts of nano-size B_4C particulates was synthesized using the disintegrated melt deposition technique followed by hot extrusion. Microstructural characterization of the developed E21-B_4C composites revealed refined grains with the progressive addition of boron carbide nanoparticles. The evaluation of mechanical properties indicated a significant improvement in the yield strength of the nanocomposites under compressive loading. Further, the E21-2.5B_4C nanocomposites exhibited the best damping characteristics, highest young's modulus, and highest resistance to ignition, thus featuring all the characteristics of a material suitable for several aircraft applications besides the currently allowed seat frames. The superior mechanical properties of the E21-B_4C nanocomposites are attributed to the refined grain sizes, uniform distribution of the nanoparticles, and the thermal insulating effects of nano-B_4C particles.

Keywords: Elektron 21; boron carbide nanoparticles; nanocomposites; properties

---

1. Introduction

With globalization, modern transport has evolved to play an integral part of the global economy; with the net worth of the automotive industry at 2 trillion USD [1] and the aerospace industry at 2.7 trillion USD [2]. However, it is also a significant source of pollution with approximately 13 wt.% of overall greenhouse gas and 25 wt.% of CO_2 emissions from the fossil fuel combustion [3]. The global aviation industry generates around 2 wt.% of all human-induced emissions and 12 wt.% of CO_2 emissions from all transport sources. As the number of passengers traveling by air is projected to increase from 3 billion in 2012 to 16 billion in 2050 and the nitrogen oxide NO_x emissions from burning jet fuel are expected to double before 2020, there is a concern about the detrimental effect on the environment. To reduce the negative impact, the Air Transport Action Group has set up several ambitious targets to become carbon neutral by 2020, while improving the fuel efficiency by 1.5 wt.% per year and halving carbon emissions by 2050 compared to 2005 levels [4]. In order to meet the environmental responsibilities, development of new structural materials for aerospace applications is driven by the need for light weight components in both cast and wrought forms. In this regard, Mg-based alloys and composites exhibit excellent specific strength and damping capacity with their very low density, and are promising materials for aerospace applications in view of light weighting the aircrafts. Over the recent years, passenger seats on-board airlines served to gain the most attention as they offer significant opportunities for weight reduction through usage of magnesium-based materials.

Until 2015, there was a ban by the Federal Aviation Administration (FAA) in the use of magnesium in aviation due to safety concerns stemming from the ignition susceptibilities of magnesium. However, as of today, Elektron® 43 (WE43) and Elektron® 21 (E21) are the only magnesium alloys that have met the cited performance requirements by passing extensive flammability tests conducted by the FAA, including seven full-scale aircraft interior tests [5]. These alloys are heavily rare earth dominated with elements such as Neodymium, Gadolinium, and other elements such as Zinc and Zirconium. These added alloying elements either dissolve in the matrix or form secondary phases with Mg, which play an extremely influential role in deciding the performance of the alloy [6,7]. With the above mentioned combination of elements, magnesium offers a combination of good castability, mechanical properties, and corrosion resistance and, hence, can be rendered useful in both civil and military aircraft and also in the automobile (motorsport) industry [8,9]. To extend their applications to other aircraft components apart from seat frames, such as gearbox housing casings, etc., improvement of strength and high temperature properties such as ignition resistance is crucial [10]. Addition of ceramic and metallic particulates as reinforcements is one of the methods to improve magnesium's property profile and performance, especially high temperature properties due to their thermal stabilities [11,12]. B_4C nanoparticles were demonstrated to be promising as reinforcements owing to their role in improving the strengths of magnesium [13]. Boron carbide is the third hardest material known to man, after diamond and cubic boron nitride, and thus the addition of boron carbide would significantly influence the properties of magnesium, which is an inherently softer matrix. Further, addition of boron carbide to a matrix with secondary phases would change the distribution of secondary phases. This is because of the difference in the sizes of the nanoparticles and the secondary phases leading to an inhomogeneity in the processing of the materials [14]. Hence, this work is aimed at understanding the influence of boron carbide on the property profile of the E21 magnesium alloy.

2. Materials and Methods

2.1. Primary Processing

E21 alloy and E21-B_4C (1.5 and 2.5 wt.%) nanocomposites were synthesized utilizing disintegrated melt deposition technique (DMD) [15]. E21 alloy procured in the form of chunks from Luxfer MEL Technologies (formerly known as Magnesium Elektron) and B_4C nanoparticles of ~50 nm size procured from Nabond, Hong Kong, were added in sandwich fashion and superheated to 750 °C, under an inert argon gas atmosphere, within a graphite crucible using a resistance heating furnace. The crucible was equipped with a plug and nozzle for bottom pouring. Upon reaching the temperature (750 °C), the melt was stirred at 450 rpm for 5 min utilizing a stirrer coated with ZIRTEX 25 to avoid iron contamination. After stirring, the plug of the crucible was pulled, and the molten metal was down poured into the mold. Before entering the mold, the molten metal was disintegrated by two jets of argon gas, with a flow rate maintained at 25 lpm. Following solidification, an ingot of 40 mm diameter was obtained. Three materials were cast in total: (i) Elektron 21 alloy; (ii) Elektron 21 with B_4C nanoparticles of approximately 50 nm added at 1.5 wt.%; and (iii) 2.5 wt.%, respectively, using this method. The composition of the final materials is given in Table 1.

Table 1. Nominal elemental composition of the materials in the study supplied by Luxfer MEL Technologies.

Material (wt.)	Nd	Gd	Zn	Zr	B4C	Mg
E21 Alloy	2.8	1.4	0.3	0.5	–	Bal.
E21-1.5B4C Nanocomposite	2.8	1.4	0.3	0.5	1.5	Bal.
E21-2.5B4C Nanocomposite	2.8	1.4	0.3	0.5	2.5	Bal.

2.2. Secondary Processing

The ingots were machined to billets with a length of 50 mm before they were sent to the extrusion process. The billets were soaked for an hour at 450 °C and extruded at 400 °C at a 20.25:1 extrusion ratio to obtain rods of 8 mm diameter.

2.3. Testing and Characterization

2.3.1. Density and Porosity Measurements

Using the rule of mixtures, the theoretical densities of the synthesized Mg materials were calculated [16]. Experimental densities of the materials were calculated using the gas pycnometer. The samples were placed in the gas pycnometer and sealed. Helium gas was then released into the chamber to measure the samples' experimental density. The porosity of the samples was calculated using the theoretical and experimental densities, under the presumption that the differential value arising between the theoretical and experimental densities is due to the porosity entrapped in the materials. The formula used to calculate the porosity is given below:

$$P = \frac{(\rho_{th} - \rho_e)}{(\rho_{th} - \rho_{air})} \times 100, \tag{1}$$

where P represents the porosity (in%), ρ represents density (g/cc), 'th' represents theoretical, and 'e' represents experimental.

2.3.2. Microstructural Characterization

The samples were ground and polished to remove any deformations or scratches on their surfaces, and etched to reveal grain boundaries. The etchant used for these materials was a solution containing 20 mL of acetic acid, 1 mL of HNO_3, 60 mL of ethylene glycol, and 20 mL of water. An optical microscope Leica DM2500 M was used to observe the grain characteristics. After the images were obtained, MATLAB software was used to calculate the average grain size of the materials. Further, the presence and distribution of the intermetallic phases and nanoparticles were studied using a JEOL JSM-6010PLUS/LV Scanning Electron Microscope (SEM), Peabody, MA, USA. Elemental analysis was performed using energy dispersion spectroscopy (EDS).

2.3.3. X-Ray Diffraction Studies

The XRD studies were conducted on the longitudinal section of the extruded samples using an automated Shimadzu lab-X XRD-6000 diffractometer (Kyoto, Japan). The samples were exposed to Cu Kα radiation of wavelength λ = 1.5418 Å with a scan speed of 2°/min and a scanning range of 20° to 65°. The bragg angles, intensity peaks, and the values of the interplanar spacing, d, obtained were subsequently matched with the standard values of Mg, Zr, Gd, Nd, Zn, B_4C, and related phases. Furthermore, the basal plane orientation of Mg-based rare earth alloys was analyzed from the XRD peaks obtained at 2θ = 34°.

2.3.4. Damping Capacity and Young's Modulus

Samples of 7.5 mm diameter and 60 mm length were subjected to impulse excitation to measure their damping characteristics (Damping Capacity, Loss Rate, Frequency and Young's modulus). The vibrational damping capacity of the materials was determined by using the response frequency damping analyzer (RFDA) from IMCE Belgium. The vibration signal was recorded in terms of amplitude vs. time. The attenuation coefficient was calculated with the help of GetData graph digitizer software, which allowed for a qualitative analysis of the damping capacity of the material.

2.3.5. Compression Testing

The compressive tests were carried out in accordance with ASTM test method E9-09 on the MTS 810 testing machine at ambient temperature, using a strain rate of 0.01 min^{-1}.

The samples with a length to diameter ratio of 1.5 were machined from the extruded rods with a diameter of 7 mm. Key mechanical properties such as 0.2% offset yield strength (0.2% YS), ultimate compressive strength (UCS), failure strain (FS), and energy absorbed (EA) were extracted from the graphs. A minimum of 5 tests were conducted to ensure repeatability.

2.3.6. Ignition Temperature Determination

The ignition temperatures of the materials were determined using a Thermo Gravimetric Analyzer (TGA). Samples of dimensions $2 \times 2 \times 1$ mm^3 were placed in purified air with a flow rate of 50 mL/min. They were heated from 30 to 1000 °C at a heating rate of 10 °C/min. The ignition temperature was recorded at a point where there was a rapid increase in the mass of the sample due to sharp oxidation upon ignition. The temperature rate was restored to the set-value after the sample burnt out. The crucible was taken out immediately after the test to prevent overflow of the oxidized power from the sample and contamination of TGA.

3. Results and Discussion

3.1. Structural Characterization

The addition of nanoparticles changes the density of the materials. The results of density and porosity measurements of as-extruded E21 magnesium alloy and E21-B$_4$C (1.5% and 2.5%) magnesium nanocomposites are shown in Figure 1. It is observed that with the addition of denser B$_4$C (density: 2.52 g/cc) to E21 alloy (density: 1.8 g/cc), the density of the E21 alloy increases. However, since the density of boron carbide is low compared to other ceramic nanoparticles, the overall density is maintained to be less than 1.81 g/cc for the nanocomposites, which is far less than aluminum-based bulk materials and only 4% higher than that of pure Mg. It can also be seen from the Figure 1 that the porosity (%) of the nanocomposites is higher than that of the alloy. However, since all the materials exhibited porosity less than 0.6%, it is implied that near dense materials have been fabricated using the DMD casting technique and the effects of porosity on the properties of the alloy and nanocomposites can be considered negligible.

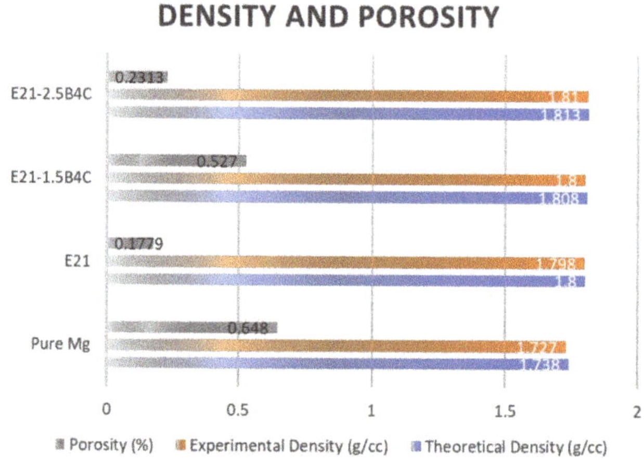

Figure 1. Density and porosity results of the as-extruded materials studied in this work.

The addition of B$_4$C nanoparticles to the E21 alloy disturbs the microstructural homogeneity of the alloy. This is because, in conjunction to the change in density, the intrinsic structure of the nanocomposites differs in terms of the grain characteristics, secondary phase distribution, and nanoparticle dispersion as compared to that of the alloy. Figure 2a–c shows the grain characteristics of the alloy and nanocomposites. Based on visual inspection and careful image analysis and quantification, it is estimated that the mean grain sizes of the nanocomposites were smaller than that of the E21 alloy. The average grain sizes of the materials are given in the insets in Figure 2, with E21 alloy exhibiting the highest average grain size. The average grain sizes are 29% and 27% lower in E21-B$_4$C and E21-2.5B$_4$C nanocomposites, respectively, compared to that of the parent material E21 alloy. Further, an interesting observation is that the difference in grain size between the two nanocomposites is very minimal.

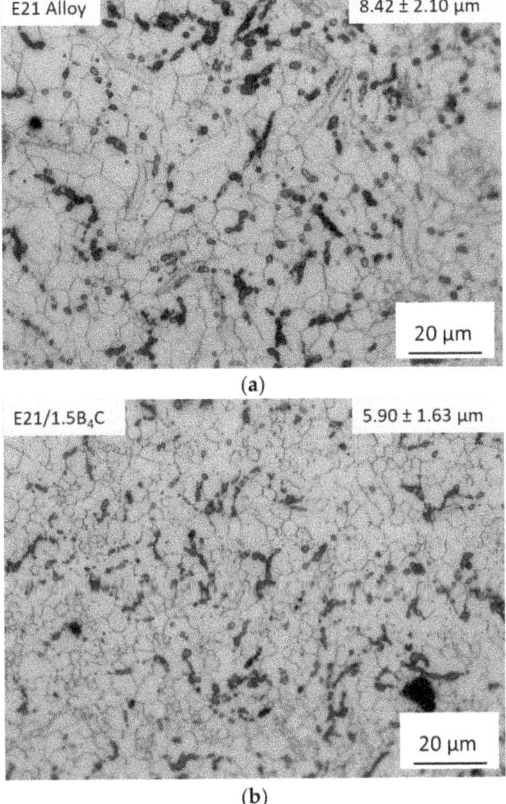

(a)

(b)

Figure 2. *Cont.*

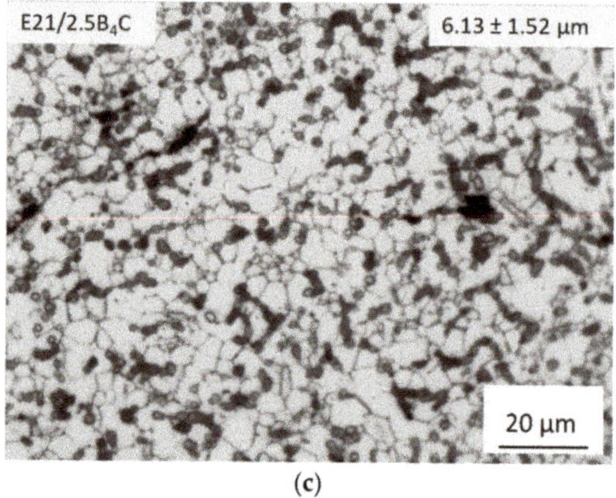

(c)

Figure 2. Optical micrographs of (**a**) E21 alloy; (**b**) E21-1.5B$_4$C nanocomposite; and (**c**) E21-2.5B$_4$C nanocomposites. The average grain sizes are given in the insets in the images.

Apart from the grain sizes, other significant changes in the microstructure with the incorporation of B$_4$C nanoparticles are the distribution of secondary phases and nanoparticles. Secondary phases of Mg-RE (Mg-Nd and Mg-Nd-Gd) and Zr-rich phases are seen in the SEM image of the E21 alloy, as given in Figure 3a and as confirmed by EDS point analysis. Zinc is found to be dissolved in the matrix owing to its good solid solubility in Mg. From Figure 2 it is evident that the secondary phases are found predominantly at the grain boundaries. With the addition of B$_4$C nanoparticles, apart from the previously discussed Mg-RE phases and Zr phases, the presence of B$_4$C particles were also observed. Although, the presence of B$_4$C nanoparticles cannot be assertively confirmed by SEM + EDS, due to the limitation of the EDS to detect elements with lower atomic numbers [17], a few boron carbide nanoparticles were identified based on the EDS point analysis as seen in Figure 3b. Based on Figure 3b, it is observed that the nanoparticles are found in the matrix. However, due to the large volume fraction of the secondary phases, the nanoparticles can also segregate towards the secondary phases [18]; however, this needs further confirmation with the use of high resolution TEM. Further, XRD was used to confirm the presence of B$_4$C nanoparticles as well as to identify the type of secondary phases. From the XRD spectra in Figure 3c, the presence of B$_4$C is evident in the nanocomposites with peaks corresponding to B$_4$C at 26° diffraction angles. In addition, the intensity of this peak is seen to increase from E21-1.5B$_4$C to E21-2.5B$_4$C due to the increased content of B$_4$C particles. Further, XRD results also confirmed the presence of the phases Mg$_3$RE and Mg$_{41}$RE$_5$, which correspond to Mg$_3$Nd/Gd and Mg$_{41}$Nd$_5$/Gd$_5$, respectively. Zn and Zr were not detected by XRD, as Zn is thought to be dissolved in magnesium, while Zr's percentage in Elektron 21 is too little and, hence, it is difficult for the filtered X-ray to detect the phase when the volume percentage of the phase present in the alloys is less than 2% [19].

Thus, from the microstructural analysis, a few distinct observations can be made: (i) Zn is present as part of the solid solution of magnesium matrix; (ii) distinct secondary phases of Mg-Nd and Mg-Gd are seen in the matrix and the different secondary phases seem to co-exist; (iii) Zr particles are seen in the Mg matrix; (iv) Mg-Nd phase had the highest concentration amongst all the other phases owing to the higher concentration of Nd in the E21 alloy [20]; (v) B$_4$C nanoparticles are present in the matrix and the nanoparticles aided the secondary phases in the grain refinement of the material due to the observation of grain refinement in the nanocomposites with the aid of nanoparticles.

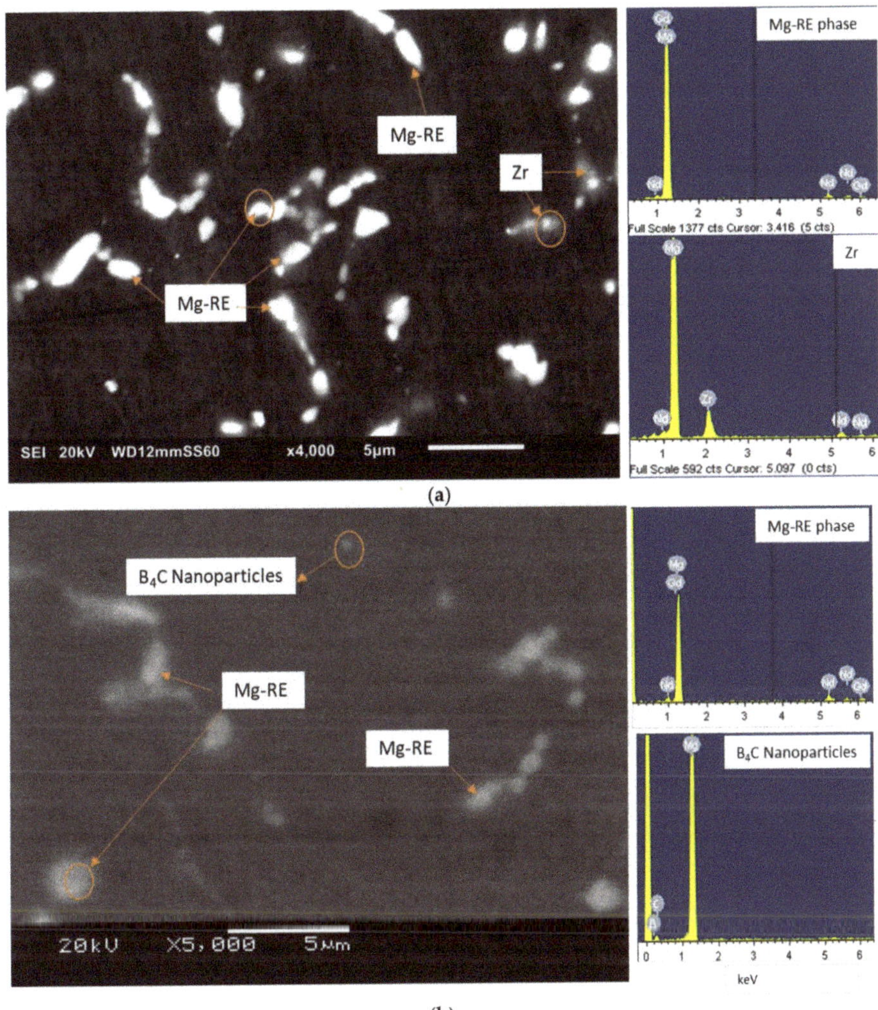

(**a**)

(**b**)

Figure 3. *Cont.*

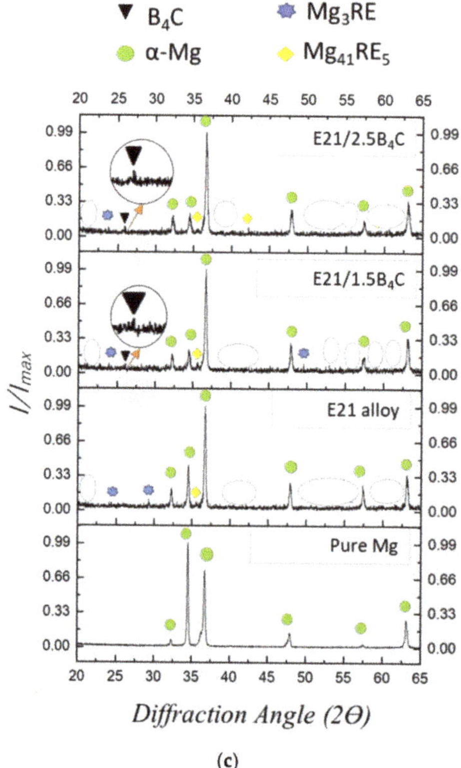

(c)

Figure 3. Scanning electron micrographs with energy dispersive spectroscopy (EDS) analysis of (**a**) E21 alloy; (**b**) microstructure of E21-1.5B$_4$C nanocomposite with EDS point analysis of boron carbide nanoparticle and Mg-RE (Mg-Nd-Gd) phase; and (**c**) X-ray diffraction spectra of the materials taken on the longitudinal section of the samples.

From the microstructural analysis of secondary phases, nanoparticles, and grain sizes, the results confirm that the nanoparticles contributed to the further grain refining of the material. This is in contrary to the previously reported alloy nanocomposites, where marginal grain coarsening was observed when nanoparticles were added to magnesium-based alloys [14,21]. In previous works, it was reported that the nanoparticles, in certain magnesium matrices, alter the mechanism of dynamic recrystallization, from Zener pinning to localized particle stimulated nucleation, causing an inhomogeneity and; therefore, a bimodal grain size across the nanocomposite [14]. However, in this case, the nanoparticles aided the secondary phases in pinning the grain boundaries and refine the grains. It is proposed that this could be due to the segregation of the secondary phases at the liquid–B$_4$C nanoparticle interface [22], thereby leading to the secondary phase and nanoparticles complementing each other in the process of grain refinement.

3.2. Property Profile

3.2.1. Mechanical Properties

The compressive results of the developed Mg alloys are shown in Figure 4, where the representative engineering stress–strain curves of the E21 alloy and its nanocomposites under compression are depicted. E21 alloy is known for its good mechanical properties exhibiting a good strength and moderate ductility.

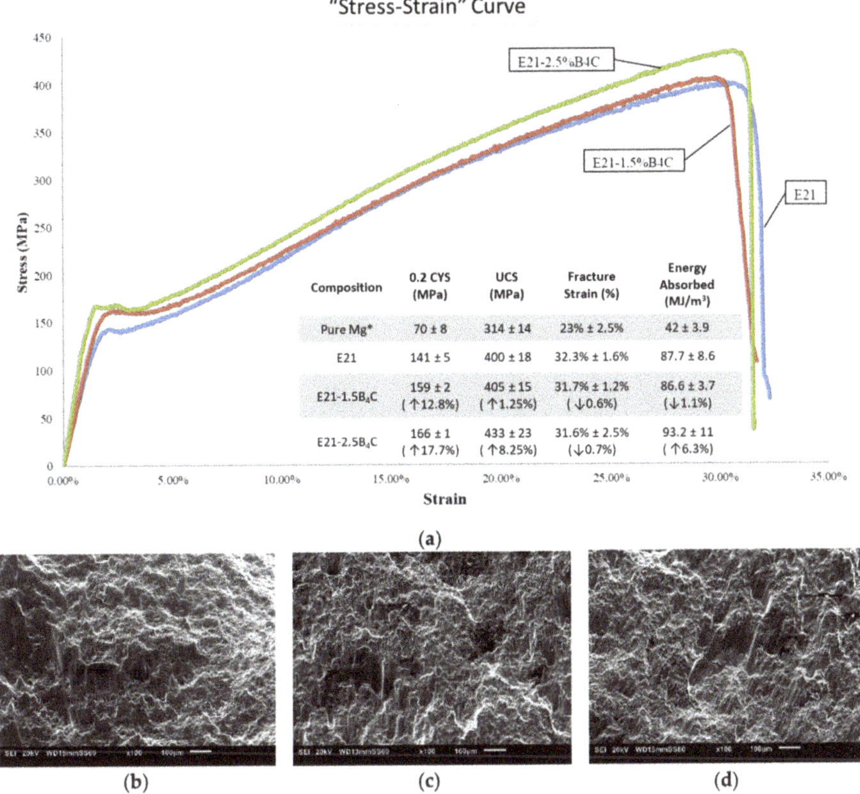

Composition	0.2 CYS (MPa)	UCS (MPa)	Fracture Strain (%)	Energy Absorbed (MJ/m³)
Pure Mg*	70 ± 8	314 ± 14	23% ± 2.5%	42 ± 3.9
E21	141 ± 5	400 ± 18	32.3% ± 1.6%	87.7 ± 8.6
E21-1.5B$_4$C	159 ± 2 (↑12.8%)	405 ± 15 (↑1.25%)	31.7% ± 1.2% (↓0.6%)	86.6 ± 3.7 (↓1.1%)
E21-2.5B$_4$C	166 ± 1 (↑17.7%)	433 ± 23 (↑8.25%)	31.6% ± 2.5% (↓0.7%)	93.2 ± 11 (↑6.3%)

(a)

(b)   (c)   (d)

Figure 4. (**a**) Engineering stress–strain curves of all materials tested under uniaxial compressive loading; (**b**) fractography of the Elektron21 (E21) sample fractured under compression; (**c**) fractography of E21-1.5B$_4$C sample fractured under compression; and (**d**) fractograph of E21-2.5B$_4$C sample fractured under compression.

From Figure 4, it is noticeable that the E21-B$_4$C nanocomposites outperform the E21 alloy in the compressive yield strengths while maintaining the same strain to failure. The YS and UCS of E21-1.5B$_4$C increases by 12.8% and 1.25%, respectively, with respect to the E21 alloy, while the YS and UCS of E21-2.5B$_4$C increases by 17.7% and 8.25%, respectively, with the E21 alloy. Further, the total energy absorbed values are seen to be the highest in E21-2.5B$_4$C with 6.3%, as compared to that of the E21 alloy.

It is widely known that, at most low-temperatures, permanent deformation of metal comes from the movement of crystalline imperfections, known as dislocations, through the grains in the metal [23]. The addition of boron carbide nanoparticles reduced the grain sizes in E21 alloy; a change in grain size affects the yield strength due to the dislocations interacting with the grain boundary as they move. These boundaries act as obstacles, hindering the dislocation glide along the slip planes. As subsequent dislocations move along the same slip plane, the dislocations pile-up at the grain boundaries. On the contrary, a larger grain would result in more dislocations within the grain, resulting in more dislocations in the pile-up. Therefore, a lower applied stress is required to produce a stress great enough to cause the grain boundary to collapse. Furthermore, due to the presence of higher volume fraction of nanoparticles present in E21-2.5B$_4$C, the contribution to strengthening by boron carbide nanoparticles is significant compared to E21 and E21-1.5B$_4$C. In addition, the further increase in YS and UCS in

E21-2.5B$_4$C could also be due to the increased dislocation density formed due to coefficient of thermal expansion mismatch between reinforcements and matrix. This creates an excellent improvement in strength while maintaining the ductility, which makes it a viable option to be used in the aerospace sector. The samples had a very large strain-to-failure of greater than 30% and the same is observed in the fractographs given in Figure 4b–d. From the figures, it is confirmed that the amount of plastic deformation underwent by all the materials is very high and the samples typically deformed by shear. Further, there is no significant difference in the mode of deformation observed among the materials which reinforces the fact that the nanoparticles did not hamper the compressive ductility in the nanocomposites.

3.2.2. Damping Capacity and Young's Modulus

Damping capacity of the materials is presented in Figure 5 in terms of time and amplitude of the material and Table 2 with results from the resonance frequency damping analyzer (RFDA).

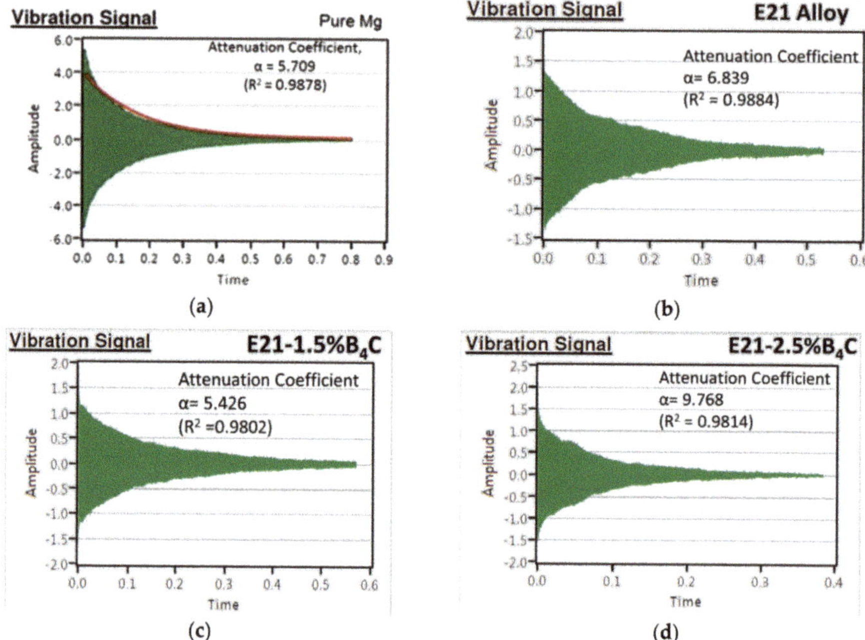

Figure 5. Damping characteristics of (**a**) Pure Mg; (**b**) Elektron 21 (E21) alloy; (**c**) E21-1.5B$_4$C nanocomposite; and (**d**) E21-2.5B$_4$C nanocomposite given in terms of amplitude vs. time.

Table 2. Results of the Resonance Frequency Damping Analyzer.

Composition	Frequency (Hz)	Loss Rate (%)	Damping Capacity	Young's Modulus (GPa)
E21 Alloy	8760.58	6.2	0.000225	48.14
E21-1.5B$_4$C Nanocomposite	8332.76	5.6	0.000213	47.18
E21-2.5B$_4$C Nanocomposite	8361.67	10.8	0.000413	52.02

The table above shows the results of frequency, damping capacity, loss rate, and Young's modulus of each composition. The damping capacity and Young's modulus are not observed to change in E21-1.5B$_4$C magnesium alloy but increase significantly in E21-2.5B$_4$C magnesium alloy. The amplitude of vibration against time for each composition can be seen in the Figure 5. The time taken for each composition to stop vibration is the qualitative measurement of the damping capacity. Among the three compositions, E21-2.5B₄C takes the least time (~0.38 s) to stop the vibration, which is far less than the time required by pure Mg to stop the vibration (~0.8 s). Currently, most of the currently used metallic materials exhibit low-damping capacity and, hence, special energy absorbers and dampeners are necessary to be incorporated into the dynamic structures [24]. This is taken more seriously in the aerospace sector, where the vibrations in the aircraft during flight can be extremely high [25]. Therefore, a large damping capacity is desirable for materials used in structures where unwanted vibrations are induced during operation, such as machine tool bases or crankshafts; especially in the aerospace sector, to reduce the chance of a crack in the components of the aircraft. This makes E21-2.5 wt.% B₄C a very suitable material, as it offers an excellent damping capacity.

Young's Modulus can also be seen to increase to as high as 52 GPa in E21-2.5B₄C. With a higher Young's modulus, the material is deemed more desirable in structural aerospace components in service for a longer duration, and works as a better replacement for other commercially used magnesium-based materials in applications requiring high stiffness.

3.2.3. Ignition Properties

The results of ignition temperature testing are shown in the schematic in Figure 6. The ignition temperature increases with the addition of B₄C nanoparticles. There is a sharp increase in ignition temperature from E21-1.5B₄C to E21-2.5B₄C. A maximum ignition temperature of 798 °C was observed in E21-2.5B₄C and the ignition temperature increased by about 57 °C (7.7%) in the nanocomposite, which indicates the effectiveness of B₄C particles in enhancing the ignition properties of magnesium.

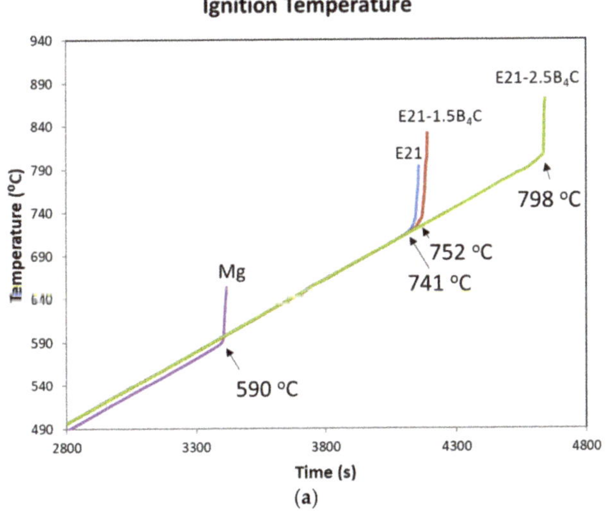

Figure 6. *Cont.*

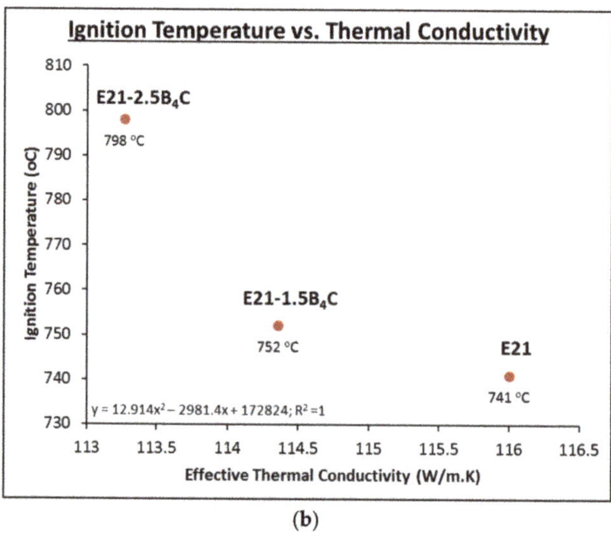

(**b**)

Figure 6. (**a**) Thermo-gravimetric analysis of materials to determine their ignition temperatures; and (**b**) ignition temperature vs. effective thermal conductivity of the materials following a polynomial fit. It is to be noted that the effective thermal conductivity is calculated using the rule of mixtures.

E21 alloy has a high ignition temperature with delayed flammability and, hence, is approved by FAA to be used in in-cabin applications [5]. This is due to the presence of a class of elements which have a positive influence on enhancing the ignition temperature. Rare earths like Nd and Gd are found in E21, raising magnesium's ignition temperature from 590 to 741 °C, as shown in Figure 6 [10]. Further, Figure 6a demonstrates that ignition temperature increases with the addition of B_4C nanoparticles. There is a sharp spike in ignition temperature from E21-1.5B_4C to E21-2.5B_4C; due to the additional 1% B_4C added in the nanocomposite. B_4C has excellent thermal properties, with a high specific heat capacity, low thermal conductivity, and a low thermal coefficient of expansion; it brings about thermal stability to E21. Figure 6b gives the relation between the thermal conductivity of the material and the ignition temperature and it has always been observed that the ignition temperature is higher for materials with lower thermal conductivity. Therefore, it is probable that the increased weight fraction of B_4C nanoparticles in E21-2.5B_4C can bring about a higher ignition temperature as it is more thermally stable as compared to the other compositions. This is essential, as it is an improvement in the thermal and ignition properties of E21 alloy, one of the few magnesium-based alloys to meet the survivability model of the Federal Aviation Administration, allowing E21 to be more reliant and suitable in the aerospace industry.

E21-2.5B_4C is an example of a nanocomposite, another emerging class of materials with extremely good mechanical properties coupled with thermal integrity owing to the presence of dimensionally stable ceramic or metallic reinforcements, which provide high mechanical strength as well as ignition resistance.

4. Conclusions

E21 has good overall properties, being one of the only few magnesium-based materials to meet the FAA's survivability requirements; good chemical resistance due to the tight control of Zn content; good mechanical, physical, and ignition properties due to the addition of Nd, Gd, and Zr. In this investigation, the effects of B_4C in E21 alloy were studied with the goal of further improving the properties of E21 alloy to be used in a wider range of applications in the aerospace sector. From the results and analysis, the following conclusions can be drawn:

(1) The presence of B_4C nanoparticles in E21 alloy did not significantly increase the density of nanocomposites but reduced the grain sizes of E21 alloy by 29%.

(2) The mechanical strength (i.e., compressive yield strength and ultimate strength of the nanocomposites, particularly E21-2.5B_4C) increased significantly without compromising on ductility.

(3) The damping capacity of E21-2.5B_4C is the highest, with the fastest time taken to stop the vibration. This makes E21-2.5B_4C a promising material to be used in the aerospace sector without the need for any special energy absorbers.

(4) The addition of B_4C nanoparticles in E21 alloys also led to the promising behavior of E21 alloy in terms of their ignition response. E21-2.5B_4C demonstrated the highest ignition temperature of 798 °C, which is about 57 °C higher than the E21 alloy, showcasing the positive role of boron carbide nanoparticles.

(5) Thus, E21-2.5B_4C nanocomposites with boron carbide nanoparticles areca desirable and suitable material for the aerospace industry to be used in in-cabin applications, as well as for gearbox housing applications.

Author Contributions: Conceptualization and design of experiments, S.T. and M.G.; performed the experiments and analyzed the data, S.T. and N.J.Y.; material selection and ideation, S.T., M.H. and M.G.; writing—original draft preparation, S.T.; writing—review and editing, S.T. and M.G.; project administration, S.T.; funding acquisition, M.G.

Funding: This research was funded by Singapore Ministry of Education (MOE) Tier 2, grant number R265-000-622-112.

Acknowledgments: The authors thank Jurami Bin Madon for his assistance with extrusion and Ng Hong Wei for his assistance with experiments on TGA.

Conflicts of Interest: The authors declare no conflict of interest. The funders had no role in the design of the study; in the collection, analyses, or interpretation of data; in the writing of the manuscript, or in the decision to publish the results.

References

1. Safehaven.com. New Tech Could Transform the $2 Trillion Auto Industry. Available online: https://www.prnewswire.com/news-releases/new-tech-could-transform-the-2-trillion-auto-industry-673561583.html (accessed on 19 August 2019).

2. Arnaldo Valdés, R.M.; Burmaoglu, S.; Tucci, V.; Braga da Costa Campos, L.M.; Mattera, L.; Gomez Comendador, V.F. Flight path 2050 and acare goals for maintaining and extending industrial leadership in aviation: A map of the aviation technology space. *Sustainability* **2019**, *11*, 2065. [CrossRef]

3. Perera, F. Pollution from fossil-fuel combustion is the leading environmental threat to global pediatric health and equity: Solutions exist. *Int. J. Environ. Res. Public Health* **2017**, *15*, 16. [CrossRef] [PubMed]

4. Czerwinski, F. Controlling the ignition and flammability of magnesium for aerospace applications. *Corros. Sci.* **2014**, *86*, 1–16. [CrossRef]

5. Marker, T.R. *Development of a Laboratory Scale Flammability Test for Magnesium Alloys used in Aircraft Seat Construction*; Federal Aviation Administration Technical Center: New Jersey, NJ, USA, 2014.

6. Lu, Y.; Bradshaw, A.R.; Chiu, Y.L.; Jones, I.P. Effects of secondary phase and grain size on the corrosion of biodegradable mg–zn–ca alloys. *Mater. Sci. Eng. C* **2015**, *48*, 480–486. [CrossRef] [PubMed]

7. Yang, W.; Tekumalla, S.; Gupta, M. Cumulative effect of strength enhancer—Lanthanum and ductility enhancer—Cerium on mechanical response of magnesium. *Metals* **2017**, *7*, 241. [CrossRef]

8. Saboori, A.; Padovano, E.; Pavese, M.; Badini, C. Novel magnesium elektron21-aln nanocomposites produced by ultrasound-assisted casting; microstructure, thermal and electrical conductivity. *Materials (Basel)* **2017**, *11*, 27. [CrossRef] [PubMed]

9. Tekumalla, S.; Seetharaman, S.; Almajid, A.; Gupta, M. Mechanical properties of magnesium-rare earth alloy systems: A review. *Metals* **2015**, *5*, 1–39. [CrossRef]

10. Tekumalla, S.; Gupta, M. An insight into ignition factors and mechanisms of magnesium based materials: A review. *Mater. Des.* **2017**, *113*, 84–98. [CrossRef]

11. Tekumalla, S.; Nandigam, Y.; Bibhanshu, N.; Rajashekara, S.; Yang, C.; Suwas, S.; Gupta, M. A strong and deformable in-situ magnesium nanocomposite igniting above 1000 °C. *Sci. Rep.* **2018**, *8*, 7038. [CrossRef] [PubMed]

12. Chen, Y.; Tekumalla, S.; Guo, Y.B.; Gupta, M. Introducing mg-4zn-3gd-1ca/zno nanocomposite with compressive strengths matching/exceeding that of mild steel. *Sci. Rep.* **2016**, *6*, 32395. [CrossRef] [PubMed]

13. Sankaranarayanan, S.; Sabat, R.K.; Jayalakshmi, S.; Suwas, S.; Gupta, M. Effect of nanoscale boron carbide particle addition on the microstructural evolution and mechanical response of pure magnesium. *Mater. Des. (1980–2015)* **2014**, *56*, 428–436. [CrossRef]

14. Tekumalla, S.; Bibhanshu, N.; Suwas, S.; Gupta, M. Superior ductility in magnesium alloy-based nanocomposites: The crucial role of texture induced by nanoparticles. *J. Mater. Sci.* **2019**, *54*, 8711–8718. [CrossRef]

15. Gupta, M.; Wong, W.L.E. Magnesium-based nanocomposites: Lightweight materials of the future. *Mater. Charact.* **2015**, *105*, 30–46. [CrossRef]

16. Chawla, K.K. *Composite Materials: Science and Engineering*; Springer: New York, NY, USA, 2012.

17. Berlin, J. Analysis of boron with energy dispersive x-ray spectrometry. *Imaging Microsc.* **2011**, *13*, 19–21.

18. Chen, Y.; Tekumalla, S.; Guo, Y.B.; Shabadi, R.; Shim, V.P.W.; Gupta, M. The dynamic compressive response of a high-strength magnesium alloy and its nanocomposite. *Mater. Sci. Eng. A* **2017**, *702*, 65–72. [CrossRef]

19. Cullity, B.D. *Elements of X Ray Diffraction-Scholar's Choice Edition*; Scholar's Choice: Wolcott, NY, USA, 2015.

20. Kiełbus, A.; Rzychoń, T.; Lityńska-Dobrzyńska, L.; Dercz, G. Characterization of β and mg41nd5 equilibrium phases in elektron 21 magnesium alloy after long-term annealing. *Solid State Phenom.* **2010**, *163*, 106–109. [CrossRef]

21. Tekumalla, S.; Farhan, N.; Srivatsan, T.S.; Gupta, M. Nano-zno particles' effect in improving the mechanical response of mg-3al-0.4ce alloy. *Metals* **2016**, *6*, 276. [CrossRef]

22. Nie, K.; Zhu, Z.; Deng, K.; Wang, T.; Han, J. Hot deformation behavior and processing maps of sic nanoparticles and second phase synergistically reinforced magnesium matrix composites. *Nanomaterials (Basel)* **2019**, *9*, 57. [CrossRef] [PubMed]

23. Tekumalla, S.; Shabadi, R.; Yang, C.; Seetharaman, S.; Gupta, M. Strengthening due to the in-situ evolution of ß1′ mg-zn rich phase in a zno nanoparticles introduced mg-y alloy. *Scr. Mater.* **2017**, *133*, 29–32. [CrossRef]

24. Li, B.; Lavernia, E.J. 3.23-spray forming of mmcs. In *Comprehensive Composite Materials*; Kelly, A., Zweben, C., Eds.; Pergamon: Oxford, UK, 2000; pp. 617–653.

25. Bai, J.; Wang, C.; Wan, F.; Yan, G. Review of aircraft vibration environment prediction methods. *Procedia Environ. Sci.* **2011**, *10*, 831–836. [CrossRef]

MDPI
St. Alban-Anlage 66
4052 Basel
Switzerland
Tel. +41 61 683 77 34
Fax +41 61 302 89 18
www.mdpi.com

Applied Sciences Editorial Office
E-mail: applsci@mdpi.com
www.mdpi.com/journal/applsci